► **Fukushima**

DOI: 10.1057/9781137274335

Also by David Elliott

THE CONTROL OF TECHNOLOGY (*with R. Elliott*)

MAN MADE FUTURES (*edited with N. Cross and R. Roy*)

THE POLITICS OF TECHNOLOGY (*edited with G. Boyle and R. Roy*)

THE POLITICS OF NUCLEAR POWER (*with P. Coyne, M. George and R. Lewis*)

THE LUCAS PLAN (*with H. Wainwright*)

ENTERPRISING INNOVATION (*with V. Mole*)

PRIVATISING ELECTRICITY (*with J. Roberts and T. Houghton*)

ENERGY, SOCIETY AND ENVIRONMENT

A SOLAR WORLD: Climate Change and the Green Energy Revolution

NUCLEAR OR NOT? (*editor*)

SUSTAINABLE ENERGY (*editor*)

DOI: 10.1057/9781137274335

palgrave▸pivot

Fukushima: Impacts and Implications

David Elliott
Emeritus Professor of Technology Policy,
The Open University, UK

palgrave
macmillan

DOI: 10.1057/9781137274335

First published 2013 by
PALGRAVE MACMILLAN

Palgrave Macmillan in the UK is an imprint of Macmillan Publishers Limited, registered in England, company number 785998, of Houndmills, Basingstoke, Hampshire RG21 6XS.

Palgrave Macmillan in the US is a division of St Martin's Press LLC, 175 Fifth Avenue, New York, NY 10010.

Palgrave Macmillan is the global academic imprint of the above companies and has companies and representatives throughout the world.

Palgrave® and Macmillan® are registered trademarks in the United States, the United Kingdom, Europe and other countries.

ISBN: 978-1-137-27434-2 EPUB
ISBN: 978-1-137-27433-5 PDF
ISBN: 978-1-137-27432-8 Hardback

A catalogue record for this book is available from the British Library.

A catalog record for this book is available from the Library of Congress.

www.palgrave.com/pivot

DOI: 10.1057/9781137274335

Contents

DOI: 10.1057/9781137274335

DOI: 10.1057/9781137274335

DOI: 10.1057/9781137274335

List of Tables

DOI: 10.1057/9781137274335

1
Introduction:
The Nuclear Back-Story

Abstract: *Nuclear power emerged after the Second World War and was adopted by many industrialised countries. From the 1980s onwards its development faltered, following some major accidents, the rise of public opposition and the advent of more competitive energy options, with some countries phasing out their nuclear programmes. However, in the 2000s, with concerns about climate change and energy security growing, it was beginning to recover. This book explores whether the Fukushima nuclear disaster in Japan will halt this nuclear renaissance.*

Keywords: nuclear decline; nuclear renaissance

Elliott, David. *Fukushima: Impacts and Implications.* Basingstoke: Palgrave Macmillan, 2013. DOI: 10.1057/9781137274335.

> Guided by electronics, powered by atomic energy, geared to the
> smooth effortless workings of automation, the magic carpet of our
> free economy heads for undreamed of destinations.
>
> Address to the American National Union of Manufactures (Rose, 1967)

1.1   Nuclear power: early ups and downs

Civil nuclear power emerged from the Second World War nuclear weapons programme in the US and then in the UK, France and the USSR,
which led, in the 1960s, to major civil reactor developments in the US,
Europe and the Soviet Union (on reactor types, see Section A.2 of the
appendix to this book). In the 1970s the technology spread, with, for
example, Japan, India and China developing civil nuclear programmes,
supported by the US, the Soviet Union, or in some cases Canada, France
or the UK. The US was energetic in promoting this technology under the
'atoms for peace' banner, but some saw this as part of an attempt to consolidate or expand its technological, political and economic hegemony –
as, for example, in its support for nuclear projects in the Philippines
under President Ferdinand Marcos. Some other developing countries
also took up the nuclear option, notably South Africa, Brazil, Argentina,
Mexico and South Korea.

However, in 1979 there was a major nuclear accident at the Three
Mile Island plant in the US, and this, along with the poor economics
of nuclear compared with other energy options, led to a halt in new
nuclear developments in the US. And then, following the even larger
nuclear disaster at Chernobyl in the Ukraine in 1986, many (but not all)
European countries backed off from nuclear, some of them introducing
phase-out policies.

The reversal in the fortunes of nuclear power at this stage has often
been linked, in the popular media particularly, primarily to these accidents, but the reality is more complex. Nuclear technology is expensive.
Despite claims that it would get cheaper, the economics of nuclear
power has always been an issue. Typically, although the net fuel costs
were lower, the capital cost of nuclear plants was at this stage running at
around three times that of fossil-fuelled plants, and it continued to rise
as safety requirements grew after the major accidents. In parallel, natural
gas emerged as a cheap fossil option, used in low-cost combined cycle
gas turbines. So the decline of nuclear in the late 20th century might be

DOI: 10.1057/9781137274335

seen as due to a combination of the inherent high costs, the extra cost of more safety systems and the advent of lower-cost fossil alternatives.

It might be added that public opposition to nuclear power also played a role: nuclear programmes were resisted strongly by environmental groups around the world from the 1970s onwards, with major 200,000-strong demonstrations in Europe (in Germany and Spain particularly) and in the US. The spate of major accidents consolidated this opposition, and public opinion polls suggested that opposition remained high (typically running at 70–80%) right up to the end of the century. For example, in the UK, opinion polls showed that around 75% of people were opposed to nuclear power after the Chernobyl accident. Interestingly, this opposition did not fade away. Instead it increased. By 1991, 78% of respondents to a Gallup poll either wanted 'no more nuclear plants' or for the use of nuclear power to be halted (SHE, 1994).

The impact of this opposition can be overstated: few projects were actually halted, although some may have been delayed. But other things being equal, few governments would be willing to court unpopularity by promoting nuclear power too strongly.

1.2 The nuclear renaissance

In the late 1990s and early 2000s, with climate change a growing issue, views began to change. Massive media coverage of climate issues meant that the public was increasingly concerned about the impacts of global warming, and nuclear was regularly portrayed as a way forward, with much talk of a nuclear renaissance (Nuttall, 2005). For a new generation, Chernobyl was perhaps seen as long ago and far away. This process of reassessment was no doubt helped by what some might see as an attempt to rewrite history, with reports emerging that suggested that the death rate resulting from Chernobyl was in fact much lower than had been thought earlier.

The interpretation of the epidemiological data from Chernobyl exposures remains controversial. A '10 years after' UN review suggested that around 2,500 of the 200,000 'liquidators' who were brought in to clean up the Chernobyl plant might develop cancers, as might a further 2,500 people from the immediate area. Initial studies reported about 4,000 cases of thyroid cancer in children and adolescents who were exposed at the time of the accident. However, most of the thyroid cancers were not

DOI: 10.1057/9781137274335

fatal as they were treatable by thyroid removal, albeit with a substantially reduced quality of life (UNSCEAR, 2000).

So although there were impacts, the final death rate would, it was claimed, be low, and some reports began to emphasise the speculative nature of the longer-term death estimates. Certainly there are problems of attribution, given that, for example, it can take decades for most types of solid cancer to appear and it may be hard to prove that these and other illness are directly linked to Chernobyl. While recognising that there were cases of cancer, in 2002 a UN report claimed that some of the post-Chernobyl health effects might have had psychosomatic causes or have been due to the stress resulting from over-zealous relocation of people out of the contaminated area (UNDP/UNICEF, 2002).

A 2006 report from the Chernobyl Study Group, convened by the International Atomic Energy Agency and the World Health Organization, and involving representatives from the governments of the impacted countries, was somewhat less sure. While accepting that stress was an issue, their international expert group predicted that among the 600,000 persons who received the more significant exposures (liquidators working in 1986–1987, evacuees and residents of the most 'contaminated' areas), about 4,000 extra fatal cancers might occur. Among the 5 million persons residing in less contaminated areas with lower doses, an additional 5,000 cancer deaths were predicted, although this number was said to be more speculative (IAEA, 2006). The debate continued.

Meanwhile, the nuclear industry tried to revive its market position, stressing that nuclear was a low-carbon energy source. By the late 2000s something of a global nuclear renaissance was said to be emerging, led by China and India. In addition, in the early 2010s, some EU countries were reversing their opposition to nuclear, Russia was expanding its programme and the US was looking to a new programme. Keen to expand the market further, some nuclear technology vendors also looked to South America, where Chile and Venezuela had expressed interest (Russia offering to help Venezuela), and to the Middle East, including Egypt, Saudi Arabia and the UAE. Jordon also expressed interest, and Qatar and Kuwait announced nuclear plans. Iran, of course, already had a nuclear programme, as did Israel, although the issue of civil–military links had led to major political conflicts.

Although most environmental groups around the world had opposed nuclear power strongly over the years, and continued to do so in the 2000s, in the early 2010s the nuclear lobby was encouraged by the fact

DOI: 10.1057/9781137274335

that a handful of environmentalists changed sides and backed nuclear as a way to respond to climate change (WNN, 2009).

In parallel, with renewable energy making relatively large gains across the world, the anti-nuclear view was increasingly based not so much on concerns about accidents and leaks as on economics and strategic issues. For example, it was argued that investment in nuclear would detract from the development of renewable energy and energy efficiency, which were claimed to be much more effective ways to respond to climate change (Elliott, 2010). But the safety, risk and heath issues came back onto the agenda with a vengeance in March 2011, with the Fukushima nuclear disaster.

This book surveys the impacts of this major accident on the future of energy policies around the world. Does it herald the end of the nuclear renaissance, or will the prospects for nuclear power recover, as they did, to some extent, after Chernobyl?

DOI: 10.1057/9781137274335

2
Fukushima: The Immediate Impacts

Abstract: *The major accident at the Fukushima Daiichi nuclear power plant complex in March 2011 was initiated by a very large earthquake and tsunami, which destroyed the reactor cooling systems and led to massive explosions and the release of radioactive materials. As this chapter describes, a mass evacuation was begun while the plant operators struggled to get the reactors under control. There were fears concerning radioactive contamination of food and water from the fallout, with the impacts of the disaster, in terms of public and governmental reactions, not being limited to Japan, as is explored in subsequent chapters.*

Keywords: Fukushima accident; radioactive contamination; reactor damage; TEPCO

Elliott, David. *Fukushima: Impacts and Implications.* Basingstoke: Palgrave Macmillan, 2013. DOI: 10.1057/9781137274335.

DOI: 10.1057/9781137274335

> The situation on the site was far beyond the originally estimated accident management conditions, and as a result, the expansion of the accident could not be prevented under the framework of the prepared safety measures.
>
> Fukushima plant operator (TEPCO, 2012)

2.1   The Fukushima accident

On 11 March 2011 a Richter scale 9 earthquake occurred, with its epicentre 100 km off the northeast coast of Japan. It was followed by a giant tsunami. The quake and tsunami caused a massive amount of damage and resulted in considerable loss of life. They also disabled or destroyed key parts of Tokyo Electric Power Company's (TEPCO's) coastal-sited Fukushima Daiichi nuclear power plant complex, which used US-designed boiling water reactors (BWRs).

Crucially, swamped by the 13–14 m tsunami, diesel backup generators were flooded and failed. Given that the main grid-power to the site had also failed, these generators should have been a key backup, providing electricity to run cooling pumps. Emergency batteries soon ran out of power. Although all the reactors had been shut down as soon as the quake hit, a large amount of decay heat was still being released and, as pumping failed, temperatures began to rise. It is now known that some of the reactor fuel cores melted and burnt through the inner containment.

Desperate attempts were made to provide alternative power for pumping, using seawater at one stage. But operations were hampered by the radiation from the damaged plant. As core temperatures rose, it seems that the zirconium alloy fuel cladding reacted with cooling water to produce hydrogen gas, which, reaching a critical point, led to major explosions, blasting apart the reactor outer buildings, first at Reactor 1, about five hours after the tsunami, and then, later, at Reactor 3. There was also an explosion at Reactor 4. Although the fuel melted, it remained within the structure, but the explosions scattered radioactive debris into the air, and around the site. There were also problems with some of the used fuel stores on the site: they too lost crucial cooling, and some contained relatively fresh spent fuel with high levels of activity and heat output. There were 3,400 tonnes of used fuel in seven storage pools, in addition to the 877 tonnes of fuel in the reactors.

DOI: 10.1057/9781137274335

Images of the disaster, as it unfolded, were captured and relayed world-wide by the media, with footage of the spectacular explosions being run on news loops continuously. The media also covered the evacuation process that was instigated. Eventually around 150,000 people were moved out of the area, many of them being subjected to monitoring to see whether they had been contaminated by fallout. This led to some harrowing images of children being scanned for radiation. Fortunately, it seems that levels of exposure were low, and it was hoped that the administration of iodide tablets would prevent the thyroid cancer risks that faced children at Chernobyl. However, there was increasing concern among displaced residents, outside the 20 km exclusion zone, that not enough accurate and up-to-date information was being provided by the authorities and, in the days and weeks after the accident, these concerns began to spread across the country.

So far the only deaths that have been recorded were due to the explosions and other incidents on site, put by one commentator as 7 among the first responders and plant operators, plus 14 elderly people who died during the evacuation process (Sovacool, 2011). None of these deaths were due to radiation exposure. But, inevitably, fears remain about longer-term effects.

The Japanese Health and Labour Ministry reported that nearly 100 workers at the Fukushima Daiichi site had exceeded the legal limits for radiation doses by June 2011 (see Section A.3 for details of the safety levels involved).

A report to the American Nuclear Society in June 2011 suggested that across Japan cancer deaths due to accumulated radiation exposures could not be ruled out, and might be of the order of 100 cases (Caracappa, 2011). Unfortunately, as we shall see, scientific opinion on the risk of radiation exposure is strongly divided, and that figure might prove to be an underestimate.

It is clear that this was a very serious accident. While it could have been much worse, in that there were no catastrophic releases of radioactive material from the reactors as had happened at Chernobyl, even so, for many people what Fukushima brought home was the overwhelming scale and implications of what could happen, if not this time, then the next.

An emergency evacuation plan was produced for Tokyo and other cites within 250 km of Fukushima, although it was kept secret to avoid panic in some of the world's most crowded urban areas (McNeill, 2011a). Japan's then Prime Minister, Naoto Kan, said that at one point, in his mind, he had simulated a worst-case evacuation scenario that included the 35 million people in the Tokyo metropolitan area. That would have been impossible to organize quickly, but ultimately could have been necessary if the reactor cores had exploded. Then, depending on the prevailing wind, the city could have been uninhabitable for decades. In addition, he said, 'not only would we lose up to half of our land, but spread radiation to the rest of the world. Our existence as a sovereign nation was at stake' (Sekiguchi, 2012).

Fortunately the worst was avoided. But, for most people, what actually happened was still bad enough. The minute-by-minute timeline of events as they unfolded at Fukushima, as provided by the operating company (TEPCO, 2012), makes for worrying reading: the staff rushed from crisis to crisis, often unable to decide what was happening and what it was best to do next, and constantly being exposed to radiation. The windows of the emergency response centre had to be covered with lead shielding to reduce dose rates. But key workers, reduced to a 50-strong core group, stuck with it, and, along with the emergency services and military teams sent in to help, deserve much credit for their attempts to avert an even worse crisis. The situation outside the plant was also very challenging; the main immediate problems were the lack of timely information, the lack of coordination and the disabling effects of worry and uncertainty among many of the authorities and in the wider community.

In the following section I will look briefly at the immediate aftermath of the accident and then, in subsequent chapters, at reactions around the world, before moving on to an analysis of implications for the future.

2.2　Aftermath

It is not my intention to relay full details of the accident or its immediate aftermath here, since this has been done extensively elsewhere from a range of perspectives (INPO, 2011; Kantei, 2011; TEPCO, 2012; Large, 2012). My concern here is with the wider energy policy implications. However, many controversies remain about the handling of, and responsibility for, the accident. For example, the authorities initially denied that major meltdowns had occurred, whereas they are now known to have happened early on (Large, 2011).

There is also the question of whether significant damage was done to the reactors by the earthquake, before the tsunami hit. There were reports by workers of burst pipe-work, which might mean that loss of cooling and fuel meltdown started quite early on (McNeill and Adelstein, 2011). This is a key issue for the future since, while it is possible to consider ways of protecting plants from tsunamis, such as by installing backup pumps in flood secure locations, much more radical remedial work would be needed if key parts of the plant in fact failed as a consequence of the quake. Globally, it has been estimated, 76 nuclear plants, in Japan, Taiwan, China, South Korea, India, Pakistan and the US, are near coasts at risk of tsunamis, with 17 being especially at risk. But there are claimed to be around 90 plants globally in or near seismically active areas, 34 of them in high-risk areas, including 30 in Japan and Taiwan. So it is important to know what happened at Fukushima, even given the fact that the quake was well beyond design specifications (Tamman *et al.*, 2011).

The official view was that there had been no serious damage due to the quake. The Japanese government's report for an International Atomic Energy Agency (IAEA) ministerial conference on nuclear safety commented, 'Although damage to external power supply was caused by the earthquake, no damage caused by the earthquake to systems, equipment and devices important for nuclear reactor safety has been confirmed.'

The report did provide a let-out: 'further investigation should be conducted as the detailed status remains unknown'. However, further investigation was, of course, hampered by the fact that the plants were subsequently partly wrecked by explosions and access was difficult because of the remaining radioactive contamination.

TEPCO's interim report on the accident in January 2012 said that visual inspections had been carried out after the earthquake and before the tsunami. It noted that in the case of the Unit 1 isolation condenser, inspections found no damage to vessels or piping, and no evidence of high-pressure steam leaks. Interviews with staff who checked on the Unit 3 high-pressure coolant injection system suggest it is unlikely that there were pipe ruptures (TEPCO, 2012). So, for now, there the matter rests and, although doubts continue to be raised, the official view remains that the problems were all due to the tsunami.

Another crucial but as yet unresolved question is whether some nuclear fission episodes (i.e. criticalities) occurred in the melted uranium fuel after the initial emergency shutdown. Could some of the explosions that occurred have been minor nuclear explosions, rather than hydrogen gas explosions? Moreover, some commentators have

DOI: 10.1057/9781137274335

said fission might still be occurring in the damaged reactor cores. There have been reports that traces of short-lived fission products have been detected. Post-meltdown fission reactions have not yet been confirmed, but a paper published in *Nature* in December 2011, co-written by a former Prime Minister of Japan, included this as a worrying possibility (Taira and Hatoyama, 2011)

Questions such as these will hopefully be answered in time, when access to the plant is possible. In the immediate aftermath of the accident there were more pressing issues to attend to. Much needed to be done, notably reducing the leakage of highly radioactive cooling water and stabilising the reactors and the spent fuel stores. This was hard because it was not known exactly what had happened inside the reactor cores. It took many weeks, indeed months, for the nuclear fuels to be cooled down to the point where further explosions were less likely. At the end of 2011, the Japanese authorities announced that the Fukushima Daiichi reactors had been brought to 'cold shutdown', nine months or so after the disaster. But that just meant that temperatures were lower and more stable, not zero, as a result of continued cooling and reduced melted core activity levels. Moreover, concerns about stability and control were renewed when in February 2012 temperatures in one plant briefly started rising again (although this may have been a false reading). While gradual progress was made on dealing with leaks of radioactive water, it could take a long time to deal with the remaining problems and remove the melted nuclear fuel and radioactive debris, and even longer to clear up the site. TEPCO's current plan is for final decommissioning of the site by 2041–2051.

While the crisis at the plant had been challenging enough, in the immediate aftermath of the explosions, there were also major problems outside the plant. Although the melted fuels remained within the reactor buildings, the accident led to the release of large amounts of radioactive material. That triggered the evacuation programme and decontamination of people exposed to high doses.

Subsequent reports have suggested that some of the emergency aid was poorly delivered. For example, it has been alleged that there were long bureaucratic delays in administering iodine tablets to some of those who might need them (Greenpeace, 2012). There was also, it seems, confusion about the fallout patterns, with some evacuees evidently being sent to areas that turned out to have high contamination levels. Radiation hot spots, some well outside the 20 km exclusion zone, made the situation even more difficult.

DOI: 10.1057/9781137274335

Some of the problems should no doubt have been foreseen in disaster planning. However, the disaster was off-scale: the damage to infrastructure caused by the earthquake and tsunami clearly made effective responses much harder. For example, in some areas there were shortages of water for decontamination.

In addition to the need for evacuation and emergency aid, there were major problems with wider contamination. Radioactivity levels that exceeded regulatory standards were detected in milk, crops and fish from surrounding areas. For example, in April 2011, sand lances (a species of small fish) caught off the coast of Fukushima and Ibaraki prefectures were found to have 12,500 becquerels per kilogramme (Bq/kg) of radioactive caesium, 25 times the legal limit, and 12,000 Bq/kg of radioactive iodine, 6 times the legal limit. Eleven types of vegetable grown in Fukushima prefecture were also found to be contaminated in April. For example a Japanese parsley called seri grown in Soma City contained 1,960 Bq/kg, four times the legal limit. (A becquerel is a unit of radioactivity denoting one nuclear disintegration per second; see Section A.3.)

The government acted to prevent these foodstuffs from being consumed. For example, some of the shiitake mushroom crop grown outdoors in eastern Fukushima was banned from sale in July 2011 because of radioactive contamination.

However, there were allegedly many problems with the government's handing of these issues. For example, according to a report in the *New York Times*, government inspectors declared rice in the Onami area, 35 miles from Fukushima, safe for consumption after testing just 2 of the area's 154 rice farms. But a few days later, 'a skeptical farmer in Onami, who wanted to be sure his rice was safe for a visiting grandson, had his crop tested, only to find it contained levels of cesium that exceeded the government's safety limit. In the weeks that followed, more than a dozen other farmers also found unsafe levels of cesium' (Fackler, 2012).

The *New York Times* report commented that episodes like this 'had a corrosive effect on public confidence in the food-monitoring efforts, with a growing segment of the public and even many experts coming to believe that officials have understated or even covered up the true extent of the public health risk in order to limit both the economic damage and the size of potential compensation payments'.

It quoted Mitsuhiro Fukao, an economics professor at Keio University in Tokyo: 'Since the accident, the government has tried to continue its business-as-usual approach of understating the severity of the accident

DOI: 10.1057/9781137274335

and insisting that it knows best. But the people are learning from the blogs, Twitter and Facebook that the government's food-monitoring system is simply not credible.'

With people losing confidence in government assurances, the *New York Times* report noted, 'More than a dozen radiation-testing stations, mostly operated by volunteers, have appeared across Fukushima and as far south as Tokyo, 150 miles from the plant, aiming to offer radiation monitoring that is more stringent and transparent than that of the government' (Fackler, 2012).

The government certainly faced major logistical problems. Initially there were not enough radiation detection devices to test every product from every farm, and random sampling proved inadequate because the explosions at the plant spread radioactive particles unevenly across communities, creating small local hot spots of high radioactivity. Nevertheless, prefectural officials told the *New York Times* that since the discovery of tainted rice, they had tested rice from 4,975 farms in Onami and 21 other communities, mostly in the relatively contaminated areas to the northwest of the plant. They said that the rice from about one-fifth of those farms contained caesium, though most of it at low levels. Only 30 farms were said to exceed Japan's current safety level for food contamination at that point.

However, later reports said that, in all, the ban on sales imposed by the government would affect 154 farms that produced 192 tonnes of rice annually, with radiation levels in some samples reaching 630 Bq/km, compared with the government safety limit of 500 Bq/km (*Telegraph*, 2011a). Unsurprisingly, sales of foodstuffs from the area, and other areas like it, fell.

Some of the contamination was less easy to avoid. At one stage, tap water in some prefectures was found to have unacceptably high levels of radioactivity. Most notably, for just over a day in March 2011, radioactive iodine-131 levels in Tokyo tap water exceeded the regulatory limits set for infants. Bottled water sales grew (IISS, 2011).

As in the rural areas, some residents in Tokyo, unconvinced by government reassurances that all was well, also started measuring radiation levels themselves. The Tokyo citizens' group the Radiation Defence Project, which grew out of a Facebook page, in consultation with the Yokohama-based Isotope Research Institute, collected soil samples from near their own homes and submitted them for testing. Some of the results were shocking: one sample collected under shrubs near a baseball field measured nearly 138,000 Bq/m². Of the 132 areas tested, 22 produced samples above 37,000 Bq/m², the level at which zones were considered contaminated for evacuation purposes at Chernobyl.

Hot spots are, of course, different from the sort of full-scale widespread contamination found at Chernobyl, but Kiyoshi Toda, a radiation expert at Nagasaki University's Faculty of Environmental Studies and a medical doctor, told the *New York Times*, 'Radioactive substances are entering people's bodies from the air, from the food. It's everywhere. But the government doesn't even try to inform the public how much radiation they're exposed to' (Fackler, 2012).

Some of these fears may be overstated, and some amateur measurements may have been faulty, but some of the monitoring appears to have been quite professional (Safecast, 2012). Certainly, as we shall see, the fears, and the negative reactions to nuclear power that followed, were shared by many people around the world. As a result, the Fukushima accident may have a lasting impact not just on health concerns, but also on energy policy, both in Japan and elsewhere. Indeed, this has already happened.

This is not surprising. The widely viewed spectacle of what looked like increasing helplessness in the face of continually failing systems was very sobering. It was regularly said that if this could happen in a technologically advanced country such as Japan, then who could rely on nuclear power? As TEPCO put it, 'The situation on the site was far beyond the originally estimated accident management conditions, and as a result, the expansion of the accident could not be prevented under the framework of the prepared safety measures' (TEPCO, 2012).

The issue rapidly became a global one, not least since the BWRs at Fukushima were also used elsewhere. They were built by a consortium of General Electric, Hitachi and Toshiba, and had entered service in 1971. Twenty-six similar GE–Hitachi BWRs are still operating in the US, as well as in Spain and elsewhere (see Section A.2).

Some experts see the BWR design as basically flawed. Water boils off to steam directly in the core rather than going through a separate heat exchanger loop, as in the more common pressurised water reactor (PWR). Since their unpressurised cooling water boils at a lower temperature, BWRs also need a higher coolant flow rate, more water and powerful pumps – hence the urgent need at Fukushima to provide enough water when the main and backup pumps failed.

Some commentators have said that this is mainly a Japanese problem, since few other countries, at least outside Asia, are likely to be hit by major tsunamis. However, in 2006 there was a failure in a Swedish plant's emergency backup system (Höglund, 2006). Fortunately a disaster was avoided, but it had not needed a tsunami to create a serious threat, and it has been claimed that this failure indicates that problems at nuclear plants could be more widespread (Bethge and Knauer, 2006). Moreover, although tsunamis may be relatively rare in most parts of the world, flooding is not, and it is already an issue for nuclear plants in some locations. Flooding could also become a more serious threat for the many plants around the world sited on coasts as climate change begins to have an impact.

In 1999 a tidal surge inundated the nuclear plant at Blayais near Bordeaux, which has four reactors, three of which were running at full power (Mattei *et al.*, 2001). The flood resulted in the loss of the plant's off-site power supply and knocked out several safety-related systems, including parts of the emergency core cooling system. Fortunately the reactors were shut down successfully and, although one of the backup diesel generators failed, core cooling was maintained and a disaster was avoided.

Tragically that was not the case at Fukushima. There, the lack of preparedness and the poor handling of events, coupled with what some see as the inherently risky nature of the technology, arguably made the situation even worse. Certainly Greenpeace felt able to argue, in a 2012 report on the lessons of Fukushima, that, although events were triggered by the quake and tsunami, 'it was not a natural disaster which led to the nuclear disaster at the Fukushima Daiichi plant, but the failures of the Japanese Government, regulators and the nuclear industry. The key conclusion to be drawn from the report is that this human-made nuclear disaster could be repeated at any nuclear plant in the world, putting millions at risk' (Greenpeace, 2012).

In the chapters that follow I look at reactions to Fukushima from around the world, starting with Japan. A 2009 Japanese government poll had found that 54% of respondents were uneasy about nuclear power. That figure was to rise dramatically following the accident.

DOI: 10.1057/9781137274335

3

Reactions in Japan and across Asia

Abstract: *The Fukushima disaster was claimed to be on a scale similar to that of the disaster at Chernobyl, and although uncertainties remained about the magnitude of the impacts, opposition to nuclear grew in Japan and there were major political recriminations. As this chapter describes, trust in the authorities and in the operating company, TEPCO, was undermined, and there was increasing pressure to abandon nuclear power in favour of renewable energy – a view adopted by Prime Minster Kan. Some neighbouring countries took similar stances, but, despite growing public opposition, the governments of China, India and South Korea continued to support nuclear expansion.*

Keywords: Asia-Pacific; China; Fukushima impacts; India; Naoto Kan; public reactions; radiation levels; TEPCO

Elliott, David. *Fukushima: Impacts and Implications.* Basingstoke: Palgrave Macmillan, 2013. DOI: 10.1057/9781137274335.

DOI: 10.1057/9781137274335

> Through my experience of the March 11 accident, I came to realize the risk of nuclear energy is too high. It involves technology that cannot be controlled according to our conventional concept of safety.
>
> Japanese Prime Minister Naoto Kan (2011a)

3.1   Impacts on Japan and initial responses

Japan is prone to major earthquakes, so buildings and other structures are designed accordingly. Nuclear plants likewise. The major seven-reactor, 8.2 GW Kashiwazaki-Kariwa complex in central Japan was hit by a Richter scale 6.8 earthquake in July 2007, which fortunately led to only a relatively small radioactive leak into the sea. As a result of the quake, 400 drums of low-level waste were dislodged in a store. The lids of around 40 drums became open to the air, and it seems that some radioactive gases were ventilated. A transformer unit caught fire, and there were reports of 50 other problems, including broken pipes and radioactive water leaks. All these incidents were said to be well below safety thresholds. However, all seven reactors were closed and remained offline for some time.

A review of other nuclear plants around the country was initiated, as most of Japan's 55 reactors are designed to withstand quakes of only 6.5 magnitude. (The Richter scale is non-linear: every unit increase in the scale is equivalent to a 10-fold increase in terms of energy effect.) An earlier proposal to raise the standard to above magnitude 7.1 had been shelved because of the high costs. However, there were clearly concerns about the risks. After the 2007 episode, Japan's Citizens' Nuclear Information Centre commented, 'Japan is simply too quake bound to operate nuclear plants.'

The Fukushima disaster put the issue forcefully onto the agenda, with the tsunami taking it to a new level. Measures had been taken to protect against tsunamis, but they proved to be totally inadequate for one of this scale: it was a clear case of events going well beyond the design specification.

One initial focus after the accident was on other reactors that might also be at risk. After many protests concerning the five-reactor Hamaoka complex, on the coast near an earthquake fault around 200 km from Tokyo, the operators agreed to close it while sea defences and safety upgrades were installed. A government analysis had predicted an 87%

DOI: 10.1057/9781137274335

chance of a magnitude 8 earthquake in the Tokai region within 30 years, with the risk of a major tsunami (WNN, 2011a).

Wider concerns also emerged after the Fukushima accident about potential contamination risks. I noted some examples of food- and water-related issues in Chapter 2. The Japanese government initially suggested that the accident had led to the release to air of about 10% of the radioactive material that was released to air from Chernobyl. Subsequent studies have raised this estimate and have shown that large amounts of water-based radioactivity were also released from Fukushima to the sea, much more than from Chernobyl to the nearby river course.

According to one estimate, 770,000 trillion Bq of radioactivity seeped from the plant into the sea in the week after the tsunami, more than double the initial estimate of 370,000 trillion Bq and about 20% of the official estimate of release to water from Chernobyl. It was said that 120 billion Bq of plutonium were released, plus 7.6 trillion Bq of neptunium-239, which decays to plutonium-239 (ENE News, 2011; see Section A.3 of the appendix).

A study by the Norwegian Institute for Air Research found that the accident released more total radioactive material, to air and water, than did Chernobyl, although some was in the form of xenon, which, though still dangerous, is less harmful than the long-lived caesium. Even so, the study claimed that total caesium emissions were about half the levels from Chernobyl (Brumfiel, 2011).

Clearly Fukushima was a major accident. Large public demonstrations are rare in relatively conservative Japan, but there were protests around the country. In April, 17,500 people protested across the country, and there was a 60,000-strong demonstration in Tokyo in September. A worldwide BBC GlobeScan poll, carried out between July and September 2011, found that 57% of respondents in Japan wanted no new plants, and 27% wanted all existing plants closed as soon as possible, with just 6% still being in favour of nuclear (BBC World Service, 2011).

Many local campaigns emerged. For example, parents protested at the government's plans to raise the permitted radiation levels so that school playgrounds in affected areas could be reopened (Sayonara Campaign, 2011; CNIC, 2011). Some groups took direct action. For example, rail workers went on strike to resist the reopening of the track from Hisanohama to Hirono. Their concern was that rolling stock was contaminated – it had been passing through the Fukushima area regularly. They and other trade unions, and linked political groups, have been very active, organising an

DOI: 10.1057/9781137274335

International Workers' Rally in November 2011 in Tokyo, with the slogan 'Abolish all nuclear plants now' (Doro-chibra, 2011).

Japan's beleaguered Prime Minister, Naoto Kan, faced with calls for his resignation, found it hard to cope with the scale of the issue and adopted a critical view of nuclear power: 'Through my experience of the March 11 accident, I came to realize the risk of nuclear energy is too high. It involves technology that cannot be controlled according to our conventional concept of safety.' He called for a national debate about energy options for the future:

> I think it is necessary to discuss from scratch the current basic energy plan, under which the share of nuclear energy is expected to be more than 50% in 2030, while more than 20% will come from renewable power. The past energy policy has regarded nuclear energy and fossil fuels as two major pillars in electricity. With the recent accident, I think two additional pillars are important. The first additional pillar is to add renewable energy, such as solar and wind power as well as biomass, to be one of the core energy resources. The second additional pillar is to create an energy-saving society where energy will not be used as much as it is now. I would like to add renewable energy and energy-saving as two major pillars and to exert further efforts to achieve them, while promoting safety of nuclear energy and reducing carbon dioxide from fossil fuels. Based on these thoughts, I would like to accelerate the discussion on reviewing the overall energy policy. (Kan, 2011a)

The implications were that Japan should put more emphasis on renewables such as solar, wind and geothermal, and there was talk of expanding its renewable energy market to ¥10 trillion. Certainly there seemed to be little chance of new nuclear builds. Indeed, a phase-out was more likely. Kan said Japan 'should aim for building a society that is not dependent on nuclear power'. But he wanted to reduce the use of nuclear energy 'in a planned and phased manner and aim to realise a society in the future where we can do without nuclear power stations' (Kan, 2011b).

This gradual transition view was supported in a *Mainichi* newspaper poll by 74% of respondents, while 11% called for an immediate end to nuclear power and 13% thought there was no need to alter policy (*Mainichi*, 2011).

The government established a series of emergency measures to cut energy use, and these, coupled with exhortations to consumers to use less, evidently cut energy demand by around 30% temporarily (Lewis, 2011), although later reports put the reduction at less. One report claimed

DOI: 10.1057/9781137274335

that the efficiency measures had delivered major changes in targeted areas, such as Tokyo, with reductions of 18% in peak demand in August 2011, while the actions of people at home had led to a 17% reduction in consumption (Mitchell *et al.*, 2012).

However, the overall situation was far from resolved, and there were political repercussions. With tens of thousands of people still displaced, many of them unlikely ever to be able to return home, continuing concerns about health, safety and the cost of the clean-up, and increasing recriminations about who was responsible for the situation, the political response was quite severe. Three ministers were sacked and the entire regulatory system was revamped. Kan promised to leave office as soon as the plant had been made secure, which he duly did in August 2011, even if its safety remained unclear at that point. He said he would devote himself to supporting renewable energy. Before he stood down, he told a G8 Summit meeting in May, 'We will do everything we can to make renewable energy our base form of power, overcoming hurdles of technology and cost.'

His successor, Yoshihiko Noda, who took over in September 2011, was less clear about a nuclear phase-out, although in January 2012 he said that Japan's dependence on nuclear power must be reduced to the 'maximum extent' while 'avoiding the creation of a tight electricity supply and demand' (NEI, 2012a). Controversially, the Japanese government has decided to continue exporting nuclear technology. But it has also started backing renewables strongly, wind and solar especially. A revised energy plan covering the years up to 2030 is due soon.

3.2   The alternatives to nuclear power

Before Fukushima, nuclear power was supplying 29% of Japan's electricity, and there were plans to expand that to 50%. Japan's heavy reliance on nuclear (compared with the global 14%) is a result of the fact that it has few indigenous fossil fuel resources and has to import most of its energy. Because the country consists of a small, heavily populated series of islands, the potential for wind and other land-using renewable energy technologies has sometimes been seen as limited. But Japan at one time did lead the world in solar photovoltaics (PV) development and production, and was also a pioneer, albeit on a smaller scale, in wave energy. It also played a major role in the early stages of the global negotiations on

DOI: 10.1057/9781137274335

greenhouse gas reduction, hosting the 1997 United Nations Framework Convention on Climate Change gathering in Kyoto, the city which gave its name to the first global climate change protocol.

However, banking crises and recessionary pressures weakened its economy, and Japan retrenched its earlier quite strong commitment to renewable energy. In 2009 it even opposed a replacement for the Kyoto Protocol, backing the weaker, non-binding Copenhagen Accord. Could its approach now be reversed? That would require a major policy shift.

A 2008 US embassy cable released by Wikileaks in 2011 reported outspoken criticisms of the existing approach by Lower House Diet member Taro Kono. He depicted the Japanese bureaucracy and power companies as 'continuing an outdated nuclear energy strategy, suppressing development of alternative energy, and keeping information from Diet members and the public'. He provided a specific example of how renewables had been sidelined, noting that 'there was abundant wind power available in Hokkaido that went undeveloped because the electricity company claimed it did not have sufficient grid capacity', when in fact there was 'an unused connection between the Hokkaido grid and the Honshu grid that the companies keep in reserve for unspecified emergencies' (Wikileaks, 2011).

New policies could obviously help. But how much energy could Japan get from wind and other renewables? While space is a major constraint, that is also true of the UK and Denmark, both of which have significant renewable energy programmes, with on-land wind dominating so far. Japan could certainly do more. On-land wind installation reached 2.3 GW there in 2010, compared with 4 GW in the UK and 3.5 GW in Denmark, and both the latter have since installed more capacity, increasingly offshore.

Offshore wind is an obvious choice for Japan. An early study suggested that up to 12 GW of offshore wind capacity could be installed around Japan by 2010, generating around 39 TWh p.a., about the same as was expected at the time from a planned 17 nuclear reactor expansion programme. Offshore wind technology has since moved on, with 6 MW turbines available and 10 MW machines in development. Floating wind turbines are also being developed for deeper water. These new technologies could help Japan access some 1.6 TW of offshore wind power that is now thought to be available with new technology further out. Before the Fukushima accident, there had already been calls for 25 GW onshore and 25 GW offshore, and a subsequent study suggested that much more could be obtained (Harper,

DOI: 10.1057/9781137274335

2011). Even taking account of locational constraints, the longer-term wind potential has been put at over 200 GW, on- and offshore, which is close to Japan's total present generating capacity (JWPA, 2010)

Solar PV is another key option for Japan. In 2009 the government set targets of having 28 GW of PV in place by 2020 and 53 GW by 2030, with 10% of total domestic primary energy demand to be met by solar PV by 2050. But it recently introduced a new feed-in tariff system to support more rapid expansion to 36 GW by 2020, and PV could well become a dominant energy option. Geothermal energy is another key option; already used for heating water in Japan, in some locations it can be used for electricity production. Wave energy is a possible, quite large, option. As a high-tech player, Japan has also developed some large fuel cells, and 'hydrogen economy' options are seen as increasingly important strategic industrial developments at various scales.

One intriguing large-scale hydrogen economy idea is an ambitious proposal for a gas pipeline across northeast Asia, to be fed with hydrogen gas (plus some methane) produced by power from wind and geothermal energy sources, and in particular the large wind energy potential that exists along the Aleutian Islands and Kamchatka Peninsula. A base to manage renewable energy would be built on one of the Kuril Islands between Japan and Russia, and would distribute the energy to eastern Russia and to Japan.

That pipeline is obviously some way off, but proposals for a high-voltage direct current Asian supergrid connecting the national electricity grids of Japan, Korea, China, Mongolia and Russia, linking up renewables around the whole region and enabling more effective grid balancing is now being followed up by the Japanese Renewable Energy Foundation (REW, 2012). Clearly, if renewables are to expand significantly in Japan, then to deal with the variability of the sources and to 'top up' when local inputs are low, a transnational supergrid of this sort will have to be developed, as is now planned in Europe.

How much energy can be expected from renewables in total? A 2003 report commissioned by Greenpeace, 'Energy Rich Japan: Full Renewable Energy Supply of Japan', claimed that Japan could make a full 'transition to clean, renewable energy without any sacrifice in living standards or industrial capacity' (Greenpeace, 2003). The report used 1999 energy data and showed that demand could be reduced by 50% with energy-efficient technologies that were already available, saving nearly 40% in the industrial sector, more than 50% in the residential and commercial

DOI: 10.1057/9781137274335

sectors and about 70% in the transport sector. The report then showed how renewable energy could be used to meet that new level of demand, reducing and ultimately eliminating the need for imports.

Six scenarios of how this might happen were outlined, moving up to 100% renewable energy for Japan. Starting from a basic model (Scenario One) providing more than 50% of total energy needs from domestic renewable sources, each subsequent scenario provides variations or expansions, gradually reducing the reliance on imported energy, factoring in different population projections and expected improvements in renewable generation capacity and energy efficiencies, until by Scenarios Five and Six no energy imports are required. And, of course, no nuclear either.

In a post-Fukushima update, Greenpeace claimed that Japan could switch off all nuclear plants permanently and still achieve both economic recovery and its carbon dioxide reduction goals. Its 'Advanced Energy [R]evolution' report for Japan claims that energy efficiency and rapid deployment of renewables could provide all the power Japan needs. Wind and solar capacity could, it says, be ramped up from the existing 3.5 GW to 47.2 GW by 2015, with around 1,000 new wind turbines installed each year and an increase in the current annual solar PV market by a factor of five, to supply electricity for around 20 million households. At the same time, load reduction strategies could cut energy demand by 11 GW (Greenpeace, 2011).

This may be very optimistic. But some of Japan's nuclear capacity has, in effect, phased itself out – very painfully. And the remaining plants were closed in sequence for checks. By early May 2012, all of Japan's 54 reactors had been shut down. It is still unclear when (or whether) any of them will be allowed to restart by the local municipal authorities. By contrast, all bar one of Japan's 1,742 wind turbines, including one part-offshore turbine located on a causeway, survived the quake and tsunami unscathed and have carried on generating. It will be interesting to see what path Japan takes.

The government has already indicated one key option: it is providing support from the emergency reconstruction budget for a 1 GW floating wind farm project off the coast from Fukushima, to be completed by 2020. That could be just the start. Certainly many new renewable energy technology ideas are being followed up in Japan with some urgency, including a novel ducted wind turbine design, the wind lens, which could be used offshore (Ohya and Karasudani, 2010).

DOI: 10.1057/9781137274335

However, following a new path will not be easy. As well as technical issues, there are many institutional obstacles to overcome, such as the low interest in renewables shown in the past by the Ministry of Economy, Trade and Industry. The continuing dominance of powerful companies such as TEPCO, and the strong government–corporate links, can also be seen as an issue. Although hard pressed financially by Fukushima, the Japanese government was reported to have decided to inject $11.5 billion to keep the company afloat (Onishi and Fackler, 2011). *The Times* said it had already provided a $5.6 billion loan (Smart, 2012).

Institutional change seems to be a pre-requisite not only for the successful development of a new approach to energy, but also for restoring public confidence in Japan's governance and institutions. Some critics have argued that there was clearly a need for change even before Fukushima, and that this is even more the case now, with trust at an all-time low (Carpenter, 2012).

3.3   Fukushima's legacy: a breakdown of trust

The Fukushima accident's legacy is not just about contamination or economic problems for Japan, or even the emergence of possible new directions for energy policies. As noted above, there were also political, institutional and corporate impacts. Ministers were sacked, the Prime Minister stood down and the regulatory system was revamped radically.

While these changes might be seen as an attempt to regain the much-reduced trust of the population in the leadership and in governance, the blame game did not stop there. TEPCO inevitably was in the front line of challenges for its handling of the plants before, during and after the accident. It already had a poor reputation. In 2002, it had been involved in a major scandal over the alleged falsification of nuclear power plant maintenance documents. The CEO and four other executives resigned, and TEPCO had to take 17 plants temporarily offline as a result (Becker, 2011).

Matters had evidently not improved much since then. For example, the *New York Times* reported that, a month before the Fukushima accident, 'government regulators approved a 10-year extension for the oldest of the six reactors at the power station despite warnings about its safety. The regulatory committee reviewing extensions pointed to stress cracks

DOI: 10.1057/9781137274335

in the backup diesel-powered generators at Reactor No. 1 at the Daiichi plant' (Tabuchi *et al.*, 2011). It went on:

> Several weeks after the extension was granted, the company admitted that it had failed to inspect 33 pieces of equipment related to the cooling systems, including water pumps and diesel generators, at the power station's six reactors, according to findings published on the agency's Web site shortly before the earthquake. Regulators said that 'maintenance management was inadequate' and that the 'quality of inspection was insufficient'.

It is not my intention to give a full account of the continuing debate over TEPCO's role. But, clearly, Fukushima led to many attacks on its credibility, with, for example, one former TEPCO executive who was a member of the Japan Atomic Energy Commission telling the *Wall Street Journal* that the disaster was worse than it had needed to be since, in its initial response, TEPCO 'hesitated because it tried to protect its assets' (Shirouzu *et al.*, 2011). One unresolved question is whether TEPCO at one point was planning to abandon the reactors to their fate – only to be over-ridden by Prime Minster Kan (Funabashi, 2012).

Some of the issues raised were practical. Why were the diesel pumps in basements? Why weren't they protected against floods? Why weren't there other off-site backups? But most of the issues concerned corporate integrity, with trust in the company at an all-time low.

One critic's view was that 'just as it had in previous years, TEPCO appears to have reported to the Japanese government only those things it was unable to keep secret. It was only when the shell of one of the reactor buildings exploded live on television on March 12 that the world became aware of the true extent of the accident' (Becker, 2001).

The accident might also have created wider doubts about Japan's technological and institutional credibility. A UK energy industry insider commented, 'One tenth of all the nuclear reactors in the world are in Japan. They have more developed technology and nuclear experience than any other nation and yet their industry is in meltdown' (Moss, 2011).

It may only have been symbolic, but the news that the US Navy, which had dispatched ships to provide assistance, had decided to sail away after becoming worried about contamination levels probably did not do much to bolster flagging national – or, for that matter, international – confidence.

DOI: 10.1057/9781137274335

Neither did the news of increased contamination of seawater off the coast of Japan. World Nuclear News reported levels of iodine-131 'well beyond normal regulatory limits' about 330 m from the Fukushima Daiichi plant, and later there were reports of iodine-131 levels being 3,355 times the legal limit. Levels of caesium-137 were also 'beyond limits' (WNN, 2011b). It was expected that this radioactive pollution would be rapidly diluted, as would airborne pollution, but the half-life of caesium-137 is 30 years, and given the weather patterns and currents in and around the Pacific area, the issue transcends Japan's borders.

3.4 Reactions across the Asia-Pacific region

The Fukushima accident inevitably had impacts on other countries in the Asia-Pacific region, many of which, like Japan, are threatened by tsunamis. Although seawater contamination was a concern, direct land contamination from fallout was not the main issue outside Japan. As the *Financial Times* noted, 'westerly winds pushed an estimated 79% of the caesium-137 out over the Pacific Ocean, with 19% deposited on land in Japan and just 2% ending up in other countries'. Although, on 15 March, changing wind meant that 'radioactivity was carried in a plume north-westwards with substantial fallout in rain and snow as far as 50 km from the plant', it was Japan that suffered most (Dickie and Cookson, 2011).

Nevertheless, there were worries about contamination among the public in nearby countries. For example, some schools in South Korea were shut because of fears among parents and teachers about radiation in rainfall, despite Seoul being 750 miles from Fukushima (Shears, 2011). Around the region, the media and the authorities were quick to issue reassurances that contamination levels would be low and well below safety limits (*China Post*, 2011; Brenhouse, 2011; Kasturi, 2011). Instead they focused on what they saw as the main concern, the future safety of their own nuclear plants. Most countries in the region reacted quickly. China, India, Taiwan and South Korea all immediately set up reactor safety reviews, although subsequent responses differed.

China's State Council suspended safety approvals for new plants, including those in the preliminary development stage, until new safety rules were in place, and it stepped up inspections on existing plants. Construction had been scheduled to begin on at least three new units. However, construction work on projects already under way was

DOI: 10.1057/9781137274335

continued, including work on reactors based on Westinghouse AP1000 and Areva European pressurised water reactor (EPR) designs.

China currently gets 2% of its electricity from nuclear and was planning to expand that to 4% by 2020. In addition to the existing 13 reactors, with a total capacity of over 10 GW, 32 more had been approved – another 34 GW. However, Fukushima, and the government review it triggered, may well lead to a reduced programme. An earlier review by the State Council Research Office, which makes policy recommendations to the State Council on strategic matters, had already indicated that it might be going too fast (WNN, 2011c).

China has evidently been having problems with its version of the 1.7 GW EPR. Two are being built at Taishan in China, 140 km west of Hong Kong. The problems seem similar to those at the EPRs being built in France and Finland: variable concrete quality, unqualified or inexperienced subcontractors, poor documentation, language issues. At Taishan, Unit 1 was meant to be ready in 2013, and Unit 2 in 2014, followed by two more (*Nuclear Monitor*, 2011).

China has also faced problems with rapidly deploying its re-engineered version of the Westinghouse AP1000. The State Council Research Office had called for a revised plan, reducing its ambitions for 2020 to around 100 GW, down from an initial 120 GW aspiration (WNN, 2011c). This may now well happen. The Economist Intelligence Unit suggested that the post-Fukushima halt in new plant approvals in China might 'be used as an excuse to lower targets for 2020, as already foreshadowed by articles in official publications'. Indeed, it put China's likely 2020 nuclear capacity at 63 GW, compared with the current target of 80 GW (WNN, 2011d).

By contrast, following Fukushima, China doubled its target for solar PV, aiming to get to 10 GW by 2015, and it plans to add a further 120 GW of hydro by 2015. (It already has around 170 GW.) It is also pushing ahead strongly with wind power: it has already reached 62 GW and could have a total installed wind power capacity of nearly 250 GW by 2020, according to the Chinese Wind Energy Association. Overall, it aims to get 15% of its primary energy from non-fossil sources by 2020, making the renewables programme far larger than the nuclear programme. And, longer term, expansion will be even more dramatic: for example, the national wind target is 400 GW by 2030, and 1,000 GW by 2050. Certainly the resources are there: in addition to large solar, biomass, hydro and tidal resources, China's onshore wind resource has been put at around 2,380 GW, and its offshore wind resource is around 700 GW, with

DOI: 10.1057/9781137274335

about 200 GW of that in shallow-water and inter-tidal areas, according to the Wind Energy and Solar Energy Resources Evaluation Centre, run by China's Meteorological Administration.

Nevertheless, the overall view of nuclear remained positive. Xu Gubao, from the National Energy Administration in Beijing, noted that although China had temporarily suspended plans for more nuclear plants following Fukushima, 'construction of nuclear reactors that the government has allotted the funds for would still go ahead'. He added, 'What is important is that we are making sure that the plants do not hold more uranium than they are designed for and there is ample water supply around these areas' (Dennis, 2011).

Whether that will be enough to assuage public opinion remains unclear. In the worldwide BBC GlobeScan poll carried out between July and September 2011, 42% of the Chinese respondents still supported nuclear power, but 35% wanted no new plants and 13% wanted all nuclear plants closed (BBC World Service, 2011).

South Korea is in a similar situation. It has 21 nuclear plants supplying around 40% of its electricity, and is keen to expand. It plans to raise the nuclear share to 56%, with 11 more reactors scheduled to come online in the next decade. It is also a significant supplier of nuclear technology around the world. However, like Japan, it is potentially at risk from tsunamis, and an Ipsos poll after Fukushima found that 61% of those asked were opposed to nuclear power (Ipsos, 2011a).

In India, after Fukushima, inspectors were told to review the safety of the country's 20 nuclear plants, 2 of which are of the same design as those in Japan. Prime Minister Manmohan Singh's $175 billion investment plan to double the number of reactors over the next two decades had already faced strong opposition, and that increased after Fukushima, when there were major demonstrations. One protestor was shot dead and several others were injured by police at an anti-nuclear demonstration at Jaitapur, where a large project is planned. There has also been local opposition to the two reactors planned for Kudankulam. The BBC GlobeScan poll found that 18% of those asked in India wanted no new plants, 21% wanted all the existing plants closed and only 23% supported nuclear.

Nevertheless, asked whether he thought nuclear energy still had a role in India post-Fukushima, Singh said, 'Yes, where India is concerned, yes. The thinking segment of our population certainly is supportive of nuclear energy' (WNN, 2012a). So the programme is still going ahead, with interest also being shown in the thorium option (see Section A.2).

DOI: 10.1057/9781137274335

Pakistan also seems unlikely to change its stance, with even more people (39%) than in India supporting nuclear in the BBC poll, although some of the opposition levels were higher than those in India (22% against new plants, 21% against all plants).

By contrast, Thailand decided to delay its first nuclear plant, and, following a safety review, Taiwan is planning to phase out nuclear, although slowly. No life extensions will be granted to Taipower's existing nuclear plants, but existing construction will continue. The long-term aim is said to be to make the island 'nuclear-free' (WNN, 2011e).

Following Fukushima, there were reports that Malaysia was also abandoning its nuclear plans, although it is not clear whether this will only be a delay (Aniletto, 2011). Reports also emerged that the Philippines would re-channel its £100 million nuclear budget into renewables: 77% of those asked opposed nuclear in the BBC GlobeScan poll, with 21% in favour. Opposition was also high in Indonesia, at 73%, with only 12% being in favour (BBC World Service, 2001), but the country's plans for a nuclear plant seem not to have changed, despite it being in the so-called ring of fire volcanic zone. Vietnam also decided to continue with its much more ambitious plan for 14 nuclear plants by 2030, supported by South Korea (Berger, 2011).

There had been some pressure to reverse Australia's long-standing opposition to nuclear generation, but, after Fukushima and the election of a left-leaning government, that is now very unlikely. A post-Fukushima poll found that 66% of those asked were opposed to nuclear power (Ipsos, 2011a). New Zealand similarly has a long-standing anti-nuclear policy, but, as with Australia, the vast distance separating it from Japan evidently precluded serious concern about contamination by air or sea: northern and southern hemisphere waters evidently do not mix much (Priestly, 2012). Nevertheless, as the past few years have shown, some of the Pacific rim countries and some parts of southeast Asia are in the front line for tsunamis. Australia and New Zealand are also at risk.

While the threat of tsunamis is likely to have had an effect on views on nuclear power in the southeast Asia/Pacific region, only a handful of countries within the region that did not already have established anti-nuclear positions have changed policy. Despite strong opposition to nuclear power in just about all the countries in the region for which there is data, their governments mostly remain strongly committed to nuclear. Within Asia, the major exception, of course, is Japan, and post-Fukushima developments there may begin to have an influence on the

DOI: 10.1057/9781137274335

region and more widely. Certainly Japan's former Prime Minister Kan saw Japan as a possible pathfinder, saying that 'we should aim to create a world in which people do not need to depend on nuclear energy', and adding that 'it would be ideal if Japan can become a model country for the world' (Sekiguchi, 2012).

While some countries in the region and elsewhere may say, 'It can't happen here', that view now seems to be less convincing, even, as we shall see, in countries not at risk from tsunamis.

DOI: 10.1057/9781137274335

4
Reactions in Continental Europe

Abstract: *The impact of Fukushima was dramatic in Western Europe. As this chapter describes, with opposition to nuclear rising to unprecedented levels, the German government shut all the country's older nuclear plants and decided to phase the rest out by 2022, focusing instead on a major renewable energy expansion programme. Italy voted against a proposed nuclear programme, and Belgium and Switzerland decided to phase out nuclear. A decision on a possible new Dutch reactor was postponed. The nuclear issue was a factor in the French presidential elections, leading to a partial nuclear phase-out plan. The impact was less significant in Eastern Europe, although opposition grew and Bulgaria decided not to go ahead with a new plant.*

Keywords: EU; France; Germany; Italy; nuclear phase-out; public reactions

Elliott, David. *Fukushima: Impacts and Implications.* Basingstoke: Palgrave Macmillan, 2013. DOI: 10.1057/9781137274335.

> We want to end the use of nuclear energy and reach the age of renewable energy as fast as possible.
>
> German Chancellor Angela Merkel, quoted in the *Guardian*, 9 May 2011

4.1   Europe: strong reactions in the West

The reactions to Fukushima were most obvious, large scale and immediate in Germany, with huge demonstrations across the country calling for a full nuclear phase-out. But Italy soon joined in the backlash against nuclear and, perhaps surprisingly, so did Switzerland and France, leading, as we shall see, to major policy changes. Opposition also grew to some extent in Eastern Europe, but there were few policy changes. I deal with the situation in the UK in the next chapter, in part since its response was, arguably, anomalous: some have claimed that support for nuclear increased, and certainly the government seemed unshaken by events in Japan.

The high level of opposition across most of Western Europe is perhaps not surprising. Nuclear power has always been controversial in most of Europe, including the UK, with Denmark, Ireland, Austria, Norway, Portugal and Greece all being non-nuclear. Denmark voted against nuclear power in 1985, and Austria mothballed an almost ready to run plant after Chernobyl. Sweden, Belgium and Spain had phase-out plans. But it is in Germany that anti-nuclear feeling has been highest.

4.2   Germany confirms its nuclear exit

There has been strong opposition to nuclear power in Germany since the 1970s, when there were major demonstrations against proposed new plants. Anti-nuclear and pro-renewable energy policies were at the core of the platform of the emerging Green Party, and were reinforced by the Chernobyl disaster in 1986: the fallout plume had reached Germany. Subsequently, after the Greens became part of a coalition government in 1998, a nuclear phase-out policy was established, based on limiting the life of existing plants. In parallel, Germany embarked on a major expansion of renewable energy, becoming a world leader in wind and solar power. Wind generation capacity expanded from less than 3 GW in 1998 to more than 27 GW in 2010. During the same period more than 17 GW of solar PV capacity was installed (BMU, 2011a). These projects were facilitated

DOI: 10.1057/9781137274335

by an innovative feed-in tariff support system. Around 370,000 jobs have been created in the renewable energy industry, with more expected.

With the rise of centre-right politics, and the Greens out of the coalition, Angela Merkel's government sought to soften and delay the nuclear phase-out and also started cutting back on the feed-in tariff, although there was never any suggestion of a nuclear new-build programme. But then, in March 2011, Fukushima changed the situation dramatically. With regional elections due and large demonstrations in favour of a complete and rapid nuclear phase-out, the German government immediately shut down all of Germany's oldest nuclear plants. In the event, despite its temporary nuclear moratorium, the government still did badly in the elections. A survey by GfK Marktforschung in April 2011 found that public support for nuclear, already previously very low, at around 10%, had fallen to 5%, although it was higher in the East, at 10%. These figures were confirmed by the post-Fukushima BBC GlobeScan poll, which found that 90% of respondents were opposed: 38% wanted no new nuclear, 52% wanted no nuclear at all, and just 7% supported nuclear power (BBC World Service, 2011).

The government undertook a review of its energy options, but the Deputy Environment Minister publicly stated that the eight oldest nuclear plants would stay shut down permanently and that a rapid phase-out of the remaining nine would follow (Reuters, 2011a). This policy was backed by the German Association of Energy and Water Industries, which called on the government to set everything in motion to speed up the transition towards a stable, ecologically responsible and affordable energy mix without nuclear power. The association represents about 1,800 utilities, among them the operators of the country's nuclear reactors, which, when all were running, generated 26% of Germany's electricity. The two biggest operators, E.ON AG and RWE AG, opposed the decision, but were outvoted (Reuters, 2011b).

The government formally announced at the end of May 2011 that all nuclear plants would be closed by the end of 2022, and that there would be a speedy transition to an energy system based on renewables (Spiegel Online, 2011).

4.3   Will Germany succeed?

German Environment Minister Norbert Röttgen told *Der Spiegel* he was confident that it could be done, given the rapid growth of renewables

DOI: 10.1057/9781137274335

and the potential for energy saving, but that 'everyone will have to invest in the energy turnaround. The expansion of renewable energy, the power lines it requires and the storage facilities will cost money. ... But after the investments are made, the returns will follow' (Spiegel Online, 2011).

So what is envisaged? Röttgen explained, 'First we'll have to focus on retrofitting buildings. The €460 million currently budgeted for that program won't be enough.' Second, there would be a major expansion of renewables, although he said there would be no need to cover Germany with wind farms, as some critics had suggested: 'We will achieve the biggest capacities by replacing smaller wind turbines on land with more powerful ones and by generating wind energy in the North and Baltic Seas.' He concluded, 'The events in Fukushima marked a turning point for all of us. Now we jointly support phasing out nuclear energy as quickly as possible and phasing in renewable energies.'

In 2010, 17% of Germany's electricity came from renewables, rising to over 20% in the first half of 2011, and there is potential for major expansion. In addition to backing a nuclear phase-out, the 2010 'Energy Concept' review, produced by the German Federal Ministry for the Environment, Nature Conservation and Nuclear Safety (the BMU), said that renewables could supply 35% of electricity by 2020, rising to 80% by 2050. It saw off-shore wind as a significant growth area, with 25 GW in place by 2030, along with major new bioenergy projects, such as biogas replacing imported natural gas. The review also called for primary energy consumption to be halved by 2050, via a major energy efficiency programme (BMU, 2011b). The BMU review provided the basis for the new German programme. The *Wall Street Journal* said that the report marked 'a significant shift as Germany ceases to debate whether to phase out its reactors and focuses more on how quickly and at what cost' (Radowitz, 2011).

The aim of the programme is to reduce greenhouse gas emissions by 40% from 1990 levels by 2020, 55% by 2030, 70% by 2040 and 80% by 2050. This is to be achieved by expanding renewables to supply 35% of electricity (18% of primary energy) by 2020, moving up in stages to 80% of electricity (60% of energy) by 2050. In addition, primary energy consumption will be reduced by 20% by 2020 and then in stages by up to 50% by 2050, with electricity use falling by 10% and 25% respectively. Germany's nuclear plants would all be closed, the first eight having already been shut and the rest closing in stages, the last 4 GW in 2022 (Maue, 2012).

The programme is clearly radical, but the influential German Advisory Council on the Environment has claimed that a transition to 100%

DOI: 10.1057/9781137274335

renewable electricity by 2050, rather than just 80%, is possible (SRU, 2011). Moreover, according to the Federal Environment Agency, in principle 'all of Germany's nuclear power stations could be taken offline permanently by 2017' without resulting in 'supply bottlenecks or in appreciably higher electricity prices'. Furthermore, it claimed, 'Germany's climate protection targets would not be compromised and imports of nuclear power from abroad are not necessary' (German Federal Environment Agency, 2011). The less ambitious 2022 closure date eventually chosen will clearly involve major changes, but should result in fewer problems.

Support for the programme involves more investment in renewables, including €5 billion to increase offshore wind power, financed by the German state development bank, KfW, and plans for the construction of 'electricity highways' to bring renewable power from windy northern Germany to industrial areas in the south. Some of the existing 7,800 km of high-voltage grid run by the German railways may be used for part of this. Major increases in grid integration with the rest of the EU are also planned.

To provide an overview of the way ahead, the German Federal government set up the Ethics Commission on a Safe Energy Supply to report on what should be done after Fukushima. Its 'Energy Turnaround' report was unequivocal: Germany should and could phase out nuclear within a decade and should commit to a collective effort to develop a new energy future. The report said that perceptions of nuclear risk had changed after the spectacle of an advanced industrial country being faced with a major crisis and being unable to bring it under control. There had been 'long helplessness' in the face of a disaster triggered by forces that had not been planned for, revealing a reliance on assumptions that had proven to be wrong. The risks now outweighed the benefits, making the alternatives much more attractive (Ethics Commission, 2011).

Nevertheless, the commission recognised that making the change could create major problems, and that there was potential for policy conflicts. For example, was it morally acceptable for Germany to export nuclear technology when it was closing down its own industry? On the alternatives, the Ethics Commission report said that, although renewables such as wind and solar could and should be ramped up rapidly, Germany also needed an extra 10 GW to replace the nuclear plants.

It said that, in addition to the renewables, 12 GW could come from new, and some already planned, combined heat and power projects by 2020 or perhaps earlier, 2.5 GW from biomass projects, 2.5 GW from

DOI: 10.1057/9781137274335

conventional plant and 4 GW from 'additional energy efficiency measures'. Emissions from new fossil plants would be offset using credits bought via the EU Emissions Trading System, so Germany would stay on course to meet its emission reduction targets. The report also backed a serious domestic and industrial energy efficiency programme and smart metering.

However, the main focus of the report was on the social and institutional changes that would need to be made. Germany has plenty of technical resources; the social changes will be harder, although the political will seems to be there to face the challenge.

This raises the question of whether that political will can be sustained. The 2011 regional elections in which Merkel did so badly, precipitating the full phase-out plan, did see the Greens poll 15% of the vote, and there clearly is strong support for the phase-out and for the alternative energy programme. But with the euro crisis and recession deepening, the political agenda will move on.

Certainly there has been no shortage of dire warnings about the problems that will, it is claimed, be faced following the closure of the nuclear plants. Some critics insist that Germany would have to import much more gas and use more coal, so that emissions would rise. However, Germany's carbon dioxide emissions fell by 2.2% in 2011, and it exported 4 TWh more electricity than it imported in the first half 2011, despite the nuclear closures – although exports were down from 11 TWh in 2010. Interestingly, in the winter of 2011/12 France had to import electricity from Germany, where renewable generation had increased by 19 TWh (Schaps, 2012).

The German authorities seem confident about the future. In a presentation in London in February 2012, Dr Georg Maue from the BMU commented that 'if Germany can achieve its 40% carbon reduction target by 2020, at least 500,000 additional jobs will be created, annual avoided fossil energy imports will be worth approx. €22 billion (approx. €38 billion in 2030), the national GDP will annually increase by around €20 billion/year, a SURPLUS of 34 € per reduced tonne of CO_2 equivalent will be realised in 2030 and the national debt would be some €180 billion Euro lower than it would be without climate protection measures' (Maue, 2012).

An overview from the Oeko-Institut and the WRI commented that '[t]he combination of a mix of policies (emissions trading, standards, regulations, incentives) with planning and investments in the longer-term

DOI: 10.1057/9781137274335

infrastructure is the pathway Germany has chosen' and suggested that other countries could learn from the German package 'to transition to an economically strong, low carbon economy' (Morgan and Matthes, 2011).

4.4 Reactions in Italy

Italy voted in a referendum in 1987, after the Chernobyl disaster, to close its four old existing nuclear reactors, but in 2011 the government had been pushing ahead with legislation enabling new build to start. Another national referendum had been proposed.

However, after Fukushima, with public disquiet growing, the government announced a one-year moratorium on its proposed new nuclear programme. Some anti-nuclear groups saw this as a way to deflect opposition in the national referendum, which in any case was quite constrained: it had force only if more than 50% of the electorate participated. There were even doubts as to whether the referendum would happen. Moreover, Italy's Prime Minister, Silvio Berlusconi, had said that one year would not be long enough to reassure Italians that nuclear was safe.

Public confidence in the government, already weak, was not increased when Stefano Saglia, the senior Industry Ministry official in charge of the nuclear relaunch, assured reporters that Italy could not suffer a Fukushima-style disaster because the Mediterranean did not experience tsunamis. He was apparently unaware that thousands of people were killed by a post-quake tsunami that hit the Sicilian city of Messina in 1908 in Europe's worst natural disaster in recent times.

In the event, the referendum was forced through, and in June 2011, after widespread grassroots campaigning to get the vote out, 94% voted against new nuclear on a 57% poll (Beyond Nuclear, 2011). This might be seen as, to some extent, an anti-government vote, but the opposition to nuclear appears to have been very strong: an Ipsos poll in May 2011 found that 81% of those asked opposed nuclear power (Ipsos, 2011a).

The government conceded defeat and, with Berlusconi leaving office soon after, has pursued a programme of support for renewables, solar PV in particular. Prospects are good for expansion: its solar capacity now rivals that of California. Italy is targeting 23 GW of PV by 2017. And it has already installed more total wind capacity than the (much windier) UK. Italy currently gets over 22% of its electricity from renewables.

DOI: 10.1057/9781137274335

4.5 Reactions in France

France is often seen as the prime example of a country where nuclear is dominant. It accounts for 74% of France's power generation, although some is exported, leaving about 65% for national use. However, the prospects for new nuclear began to look a little uncertain even before Fukushima. The new EPR under construction at Flamanville had run into major delays and cost overruns: it has been claimed to be €2.7 billion over budget (Thomas, 2011). Similar problems faced the EPR being built at Olkiluoto in Finland, which was five years late and was said to be more than €3 billion over budget (Reuters, 2011c).

Public opinion about nuclear had been positive up to that point. For example, according to an Ipsos opinion poll in 2002 almost 70% of French adults had 'a good opinion' of nuclear power. So it was perhaps surprising that public opposition to nuclear power began to rise, reaching 67% according to an Ipsos poll in May 2011 (Ipsos, 2011a). Subsequently, Reuters reported on an opinion poll in June 2011 that found three-quarters of interviewees wanted to withdraw from nuclear energy entirely, against 22% who backed the nuclear expansion programme; 70% were opposed to any expansion. Even more dramatically, the BBC GlobeScan poll found that 83% were opposed to nuclear, 58% to new plants and 25% to all plants, with just 15% in favour of nuclear (BBC World Service, 2011).

Against this backdrop, and with presidential elections due in 2012, the government announced immediate reviews of plant safety and a longer-term reassessment of future energy policy. This reassessment would look at all scenarios 'with total objectivity, in full transparency', including the complete phase-out of nuclear by 2050 or even 2040.

Clearly the nuclear issue was moving up the agenda, if for no other reason than that the economics of the French nuclear industry seemed to be unravelling. The share price of French reactor vendor Areva dropped by 25% in response to the German nuclear exit. It had already fallen 14% following the Japanese crisis. The results of the safety review also created problems, with EDF's shares falling a further 4.06%. Although the Autorité de Sûreté Nucléaire, the French nuclear watchdog, did not call for any plant closures, it said that EDF had to install flood-proof diesel generators and bunkered remote backup control rooms at its 19 plants, at a cost €10 billion.

This added to an existing problem. Most of EDF's plants were built during a relatively short period in the 1970s and 1980s, which means that many are now old. In January 2012, the French Court of Auditors produced a review of nuclear economics, noting that, if the current limit of a 40-year operating life were retained, 22 of the 58 reactors would have to shut by 2022. However, it felt that a replacement programme on any significant scale was 'highly unlikely or impossible', and it suggested that, if France wanted to maintain the same or a similar level of nuclear output, extending plant lifetime was probably the best option, since it delayed decommissioning expenses and the need for investment in new capacity. The Court put decommissioning and waste disposal costs at €79.4 billion, albeit spread over many years. Extending the lifetime of the existing plants would, however, also be expensive: it would require upgrades. EDF has estimated that extending the lifespan of its nuclear plants from 40 to 60 years would cost €40–50 billion over the next 30 years. The Court suggested this would add 10% to production costs (WNN, 2012b).

Clearly EDF's finances have come under increasing strain. It had always been claimed that nuclear power was economically attractive and would remain so as new technology emerged, but the Court of Auditors report found that electricity from nuclear plants in France cost €49.5/MWh, and it put the cost of electricity from EPRs such as Flamanville at €70–90/MWh. Export orders for EPR plants had been seen as a way to ensure the economic viability of the French nuclear industry, but the problems at Flamanville and Olkiluoto created uncertainty. In 2010, France failed to win the contract for four nuclear plants in the United Arab Emirates. In response President Nicolas Sarkozy ordered a report on the French nuclear industry. The Roussely report (named after François Roussely, a former president of EDF) said that, as a result of the Flamanville and Olkiluoto problems, 'the credibility of both the EPR model and the French nuclear industry's ability to build new reactors has been severely eroded' (Roussely, 2010).

The situation was compounded by the subsequent decision by Constellation Energy to pull out of the proposed Calvert Cliffs-3 EPR project in Maryland, US. This was a major blow because the finance for the EPR programme in France and Finland was in part based on loans backed by prospective income from projects such as this. EDF's sell-off of its UK power distribution assets was, it seems, part of its response to the worsening situation (Busby, 2011).

Nevertheless, EDF has been bullish, claiming that it invests 'more than €11 billion a year across the world', although it has decided to diversify its nuclear-dominated portfolio by building strong businesses in gas and coal, as well as in hydropower and renewables. The nuclear component of this portfolio may yet prove problematic. EDF is almost 85% state-owned, so it is unlikely to go bankrupt, but a study of the economics of EPRs by Professor Steve Thomas from the University of Greenwich has suggested that 'from a business point of view, the right course for EDF and Areva seems clear. They must cut their losses and abandon the EPR now' (Thomas, 2010). After Fukushima, reports were circulating that EDF might indeed ditch the current EPR design for future EU plants and go for the cheaper, simpler Franco-Chinese 1 GW Atmea.

In this uncertain climate, it is perhaps not surprising that the long-standing support for nuclear power among the French technocratic elite showed signs of strain. The new critical view was developed in *La Vérité sur le nucléaire* ('The Truth about Nuclear'), a book by Corrine Lepage, who served as Minister for the Environment in the government of conservative President Jacques Chirac. It includes damning allegations about the French nuclear industry, such as the escalating cost of Areva's Finnish reactor, which will have to be borne by French taxpayers. She suggests that exiting nuclear power, rather than penalising the economy, could in fact lead to reindustrialisation. If France developed its large renewable resources to replace nuclear power, the country would create new industries and jobs like those seen in Germany. Meanwhile, the French nuclear industry could turn its attention to the growing trend towards phasing out nuclear. Lepage says that France could become a leader in decommissioning nuclear power plants worldwide. She is currently serving as a member of the European Parliament, and in that role she has questioned France's nuclear choice, calling it a 'strategic error' of historic proportions.

The new leader of the far-right National Front, Marine Le Pen, has also said that nuclear is a 'dangerous form of energy'. However, it is not just the far right that has been critical of nuclear. It was also attacked in the run-up to the presidential election by the Socialists, who made common cause with the Greens and called for a reduction of France's dependence on nuclear energy for its electricity from 75% to 50% by 2025. This would entail the phasing out of 24 of the country's 58 reactors. Although the Socialists agreed that there would be no new plants, they would not agree to stopping the construction of the new nuclear reactor at Flamanville. However, they did agree that renewables were the main way ahead.

The Socialist leader, François Hollande, won the presidential election, although during the election in-fighting there were suggestions that the phase-out plan might be softened or delayed, with initially only one plant closure going ahead (Patel, 2012). Nevertheless, progressive closures of old plants seems inevitable, and with no new plants planned, just some

DOI: 10.1057/9781137274335

life extension for some of the newer plants, nuclear power will decline. During the election campaign, Hollande said he would order the closure of Fessenheim, the oldest French plant, before the end of his term in office. The Greens had wanted a 100% closure of all plants by 2025, along German lines, and although they were evidently willing to compromise in the electoral pact, given the new political alignments in France there is likely to be continued pressure for a rapid phase-out.

A full phase-out would, of course, be a major undertaking and would take time. Nevertheless, a 2006 Institute for Energy and Environmental Research scenario claimed that it would be possible to eliminate nuclear in France entirely by 2040, although the use of gas would expand initially (IEER, 2006).

This study was based on existing technology, and renewable technology has improved vastly since 2006, with France now rivalling the UK in wind power and also developing tidal technology. So this plan could possibly be speeded up, which would bring it into line with various new scenarios suggesting that the EU as a whole could get to almost 100% renewables by 2050 (ECF, 2010; EREC, 2010; PWC, 2010).

That initially sounds unlikely for France, which is maybe the worst case, given its large nuclear programme. However, it is worth remembering that, although France may currently generate around 74% of its electricity from nuclear, that represents only about 22% of its total primary energy use (for power, heat and transport); renewables supplied nearly 13% of its primary energy in 2010, and are now expanding rapidly.

4.6   Reactions in the rest of continental Europe

Ambitious plans for expanding renewables are now entrenched in EU thinking. The current target is to get 20% of all EU energy from renewables by 2020, and then move on to between 55% (in the lowest scenario) and 75% (in the highest scenario) by 2050 – with up to 97% of electricity then being supplied by renewables. Some EU countries are doing well – notably Denmark, which is well on its way to achieving its 2020 target of getting 30% of its energy from renewables, Portugal, which is aiming for 31%, Austria 34%, Finland 38%, Latvia 42% and Sweden 49%.

Renewables usually thrive best when there is no nuclear rival for funding, but some of these countries also have nuclear programmes, most notably Finland, but also Sweden. In 2009 Sweden reversed its

DOI: 10.1057/9781137274335

post-Chernobyl nuclear phase-out policy, but opposition rose after Fukushima, with 51% of respondents in a poll opposing nuclear power (Ipsos, 2011a). There is opposition too in Finland, but the country seems likely to continue with its nuclear expansion programme (Kojo and Litmanen, 2009), despite the financial problems it is facing with the construction of its EPR at Olkiluoto. It has introduced a tax on nuclear to help.

Belgium introduced something similar. Having been without a central government for over a year, it has now established one, with one unifying policy being to revert to the nuclear phase-out plan that had been established in 2003. The reinstated plan calls for Belgium's three oldest plants to close in 2015 and the remaining two by 2025. That closure programme has been only conditionally agreed; it will go ahead only after a review of the viability of the alternative options (Torello, 2012). But there is strong support for change: a post-Fukushima poll put opposition to nuclear at 60% (Ipsos, 2011a). Belgium is pushing hard on renewables, including offshore wind. It currently gets 55% of its power from nuclear.

Unlike anti-nuclear Denmark, which has focused strongly on renewables, the Netherlands has backed both renewables and nuclear. It has one nuclear plant, but there were proposals from the right-wing government elected in 2010 for a second one – the first time a Dutch government has backed new nuclear since Chernobyl, after which it had cancelled two nuclear projects (*Nuclear Monitor*, 2012). An internet poll in response to the new plan, carried out just before Fukushima, claimed that 49% of respondents were in favour of more nuclear power, while 37% were against. But the poll also found that 97% of respondents wanted more government investment in alternative energy sources. This may have been a response to the February 2011 decision by the new government to cut subsidies for renewables drastically, from €4 billion to €1.5 billion annually.

After Fukushima that policy, and the pro-nuclear policy, were both apparently retained, but in January 2012 Dutch utility Delta delayed a decision on whether to build the second reactor by two to three years owing to uncertainties about the investment climate, future electricity prices and possible overcapacity. In April 2012 the centre-right government collapsed, so, for the present, the future of nuclear power in the Netherlands is unclear.

Opposition to nuclear has continued in Spain, with, after Fukushima, 60% opposing nuclear in a poll (Ipsos, 2011a). There were major demonstrations, with a 40-year-old plant of the same design as Fukushima's engulfed by calls for a shutdown. However, the subsequent election of a

DOI: 10.1057/9781137274335

centre-right government may soften and slow the long-standing nuclear phase-out programme, with plant lifetimes being extended.

By contrast, and perhaps surprisingly, after Fukushima there were major anti-nuclear protests in Switzerland, including a 25,000-strong demonstration. The Swiss government, which had been planning nuclear expansion, decided to abandon its plan, so in effect nuclear will be phased out by 2035, as old plants close.

The situation in the 'New Europe' of Eastern and Central Europe is more complex. Some of these former Eastern Bloc states have, or had, old Russian nuclear plants, and some are looking to nuclear expansion, not least to avoid having to import power from Russia. In some cases it was a condition of EU entry that early-generation Soviet-era nuclear reactors be closed down, and €2.8 billion was provided by the EU to help. These closures often presented major short- and longer-term economic and political problems, not least because of the need to invest in alternative energy supplies. The issue came to a head in Lithuania in particular, but the nuclear issue is also key in Poland, Bulgaria, Romania, Slovakia, the Czech Republic and Hungary, all of which have plans for nuclear expansion, Fukushima notwithstanding.

That is not to say that there is no opposition in this region. For example, after Fukushima, in a poll in Lithuania, support for nuclear fell from 60% to 10%, and in Hungary opposition to nuclear reached 59%, although it was only 48% in Poland (Ipsos, 2011a). There has also been long-running opposition to, and problems over funding for, the proposed new plant at Belene in Bulgaria; in March 2012 it was finally decided to abandon it (Novinte, 2012).

Many of these countries are rich in renewable resources and are keen to push ahead with their use. Indeed, as one of the conditions of EU entry, they have to abide by often quite challenging targets under the EU Renewables Directive (Cook and Elliott, 2010). Whether they will have the financial and technical resources to do so, and also expand nuclear, remains unclear, especially given the impacts of the global recession, which has hit many of them hard.

Looking at the EU as a whole, our review so far has indicated that in most countries for which there is data, opposition to nuclear is high. While there have been no policy changes in the East, in the West there have been some very radical changes in direction, notably in Germany, Italy, Switzerland, Belgium and France. As we shall see, the situation in the UK is very different.

DOI: 10.1057/9781137274335

5
Reactions in the UK

Abstract: *Reactions to Fukushima in the UK were, perhaps surprisingly, muted, possibly in part because the government had won at least some support for a large nuclear expansion programme. This was despite earlier widespread opposition. As this chapter describes, the media may have played a role in shaping opinions on nuclear power, but compared with the situation in most of the rest of Europe, the UK position seems to be somewhat anomalous. Nevertheless, whether the UK's nuclear expansion programme will actually go ahead, given events in Germany and France, remains unclear.*

Keywords: BBC coverage; public acceptance; replacement programme; UK nuclear policy

Elliott, David. *Fukushima: Impacts and Implications.* Basingstoke: Palgrave Macmillan, 2013. DOI: 10.1057/9781137274335.

DOI: 10.1057/9781137274335

> The framing of nuclear as a low-carbon option over the last 10 years
> had clearly impacted public attitudes.
>
> Professor Nick Pidgeon, University of Cardiff (Pidgeon, 2011)

5.1   A muted response

Although the UK media carried overwhelming details as the Fukushima
disaster unfolded, the main official reaction in the UK, apart from sym-
pathy for those affected, seems to have been to assert that, in effect, 'It
couldn't happen here', together with apparent bemusement at the rapid
policy shifts in Germany. The Department of Energy and Climate Change
commented that 'Germany's move away from nuclear is difficult to understand. We will go forward with new nuclear in the UK' (Reuters, 2011d).

The government set up a review of nuclear plant safety, which duly
reported that all was fine, although it suggested attention be paid to the
layout of UK power plants, emergency response arrangements, mecha-
nisms for dealing with prolonged loss of power supplies and the risks
associated with flooding (HSE, 2011).

Beyond that there was little debate, apart from complaints from anti-
nuclear environmental groups and a handful of academics. That left the
ponderous Generic Design Assessment of new nuclear reactors and the
'nuclear justification' process to continue more or less unscathed. In a
parliamentary vote in 2010, 80% of MPs had backed the nuclear policy 'justification' package, and in July 2011 only 14 out of 650 MPs voted against the
government's nuclear policy as outlined in the National Policy Statements
for Energy Infrastructure (NPS, 2011). By contrast, in June 2011 the German
Parliament voted by 513 to 79 to phase out all nuclear power by 2022.

Local groups around the UK began to protest at the site level, but few
expected them to be able to resist what was seen as almost inevitable: a
major nuclear expansion, the largest in Europe, driven mostly by EDF
and E.ON. Cynics argued that the UK was now the only place in the
EU where they could build plant easily, with strong state support, in
kind if not in direct finance (Burke, 2012). To the extent that there was
any opposition, it came only when it appeared that the government was
seeking to evade the rules it had imposed that there should be no public
financial support for the proposed nuclear plants.

In many ways this passivity is strange, since the anti-nuclear movement in the UK had at one time been very strong. In 1986, following

Chernobyl, a Gallup poll found that only around 18% of those asked supported an increase in nuclear power. In the autumn of that year, the Labour Party overwhelmingly backed a 'Non-nuclear and that is final' position, as a *Guardian* headline had it at the time (Elliott, 1988). Opposition remained steady subsequently. In 1991 Gallup found that 78% of those asked either wanted 'no more nuclear plants at present' or for nuclear to be halted. Opposition continued at a high level over the next decade and beyond. In 2001 the British Market Research Bureau found that 68% of those interviewed 'did not think that nuclear power stations should be built in Britain in the next ten years', and in 2002 a national opinion poll carried out for the Energy Saving Trust found that only 10% supported nuclear power. In 2005, 79% of respondents in an ICM opinion poll for the *Guardian* said they would oppose a nuclear plant near their home.

Thereafter, support began to increase slightly, although a public opinion survey commissioned in 2005 by the Institution of Civil Engineers still found that only one in four people supported the construction of new nuclear power stations in the UK. As we shall see, after that, support continued to grow while opposition reduced, with, in some polls, support outweighing opposition.

It is worth exploring briefly how this change came about, since it clearly shaped responses to Fukushima.

5.2   Shifting views

In 1995, after a major review, the Conservative government concluded that 'providing public sector funds now for the construction of new nuclear power stations could not be justified on the grounds of wider economic benefits and would not therefore be in the best interest of either electricity consumers or tax payers' (Nuclear Review, 1995). In 1998, the incoming Labour government confirmed this view: 'at present nuclear power is too expensive to be economic for new capacity and in current circumstances it is unlikely that new proposals for building nuclear plants will come forward from commercial promoters' (Trade and Industry Select Committee, 1998).

In 2003, following a Cabinet Office review which backed renewables and energy conservation and saw nuclear as, at best, a longer-term 'insurance' option (PIU, 2002), the Labour government produced a White Paper

DOI: 10.1057/9781137274335

on Energy which said that the current economics of nuclear power 'make it an unattractive option and there are also important issues of nuclear waste to be resolved' (DTI, 2003). However, after another energy review in 2006, the government changed its mind, and commented in a new White Paper on Energy, 'new nuclear power stations would make a significant contribution to meeting our energy policy goals' (BERR, 2007).

The transition from the earlier anti-nuclear line had happened in stages, via an initial move to a policy of 'diminishing reliance', and then, with New Labour in power and concerns about energy security rising, to conditional support, with the then Prime Minister, Tony Blair, commenting in May 2006 that nuclear power was 'back on the agenda with a vengeance'.

The process of winning acceptance for the change was achieved by, arguably, clever handing of the debate by the government. With old plants scheduled to close, the industry had called for the UK to 'Replace nuclear with nuclear', and this slogan seemed to work. It was initially argued that all that was being called for was to replace the existing reactors on the same sites, as these plants reached retirement. On this basis, whereas in 2005, 79% of respondents in an ICM opinion poll for the *Guardian* said they would oppose a new nuclear plant near their home, a MORI poll in 2005 suggested that support for just *replacement* plants was growing, with 30% in favour. The sites already existed.

In 2006 the government ran a public consultation exercise on the nuclear issue. The way in which it was run was challenged by Greenpeace, which won a High Court order that led to a re-run. The Court said the review process was 'very seriously flawed' and 'procedurally unfair'. For example, it said that the information given on waste was 'not merely inadequate but also misleading'. However, the order did not derail the overall process: asked whether the High Court ruling would put on hold plans to build more nuclear power stations, Tony Blair told the BBC, 'No. This won't affect the policy at all. It'll affect the process of consultation, but not the policy' (Blair, 2006).

In the event, the second consultation exercise, in 2007, arguably proved little better. The Market Research Standards Board commented that the public consultation run for the government by Opinion Leader Research was in breach of the market research code of conduct. They said that 'information was inaccurately or misleadingly presented, or was imbalanced, which gave rise to a material risk of respondents being led towards a particular answer'. For example, participants were provided

DOI: 10.1057/9781137274335

with briefing notes on alternatives to nuclear. The one on solar said, 'In sunny countries, solar power can be used where there is no easy way to get electricity to a remote place.' It added that it was 'handy for low-power uses such as solar powered garden lights and battery chargers'. This at a time when there was 20 GW or more of solar PV installed globally.

More than 4,000 individuals and groups responded to the consultation or attended one of the events. Those attending the public meetings were asked to respond to the government's questions, the key one being 'In the context of tackling climate change and ensuring energy security, do you agree or disagree that it would be in the public interest to give energy companies the option of investing in new nuclear power stations?'

In response to the consultation, the government was able to say that reactions were overall positive: 'despite participants' clear discomfort with some of the safety and security implications and concerns about creating new nuclear waste, 44% agreed (15% strongly), 37% disagreed and 18% neither agreed nor disagreed'. By September 2008, the then Prime Minister, Gordon Brown, felt able to build on this and say, 'We made the right decision about nuclear power, I think very few people now doubt that.'

By this stage, the decision had expanded into an open-ended commitment to a large nuclear programme, well beyond just replacement, with ministers talking about nuclear ultimately supplying 40% of power. In the meantime, there were hopes that around 19 GW might come online by 2023, supplying about 30% of UK electricity.

However, in line with the government's free market policy, no specific targets were set. The government's 2011 National Policy Statements for Energy Infrastructure said that by 2025 the UK would need 113 GW of electricity generating capacity, of which at least 59 GW would have to be new capacity, with renewables at around 33 GW and 26 GW being 'for industry to determine', although it said it believed that, 'in principle, new nuclear power should be free to contribute as much as possible towards meeting the need for around 18 GW of new non-renewable capacity by 2025'. It noted that 16 GW of new nuclear capacity based on the Areva EPR and the Westinghouse AP1000 designs had been proposed by various consortia (NPS, 2011).

The nuclear industry and its supporters have talked in terms of more subsequently, with, for example, a second expansion phase, after 2030, leading to up to 25 GW of new nuclear capacity (Grimes and Nuttall, 2011). Under the Labour administration there had been talk of nuclear

DOI: 10.1057/9781137274335

providing 35–40% of UK electricity 'beyond 2030' (Wicks, 2009). A 2012 Energy Research Partnership report, produced by the UK National Nuclear Laboratory in consultation with the nuclear industry, suggested that, after the first 16 GW of new plants had been built, more could be added to bring the total to over 40 GW by 2050 (ERP, 2012). Moreover, a report from the Smith School of Enterprise and the Environment, led by Professor Sir David King, who was Chief Scientist in Tony Blair's administration, has talked of moving up to 90 GW by 2050 – nine times current capacity (Smith School, 2012).

For the initial programme, eight plants were proposed by various consortia, all on existing nuclear sites, with completion dates around, or soon after, 2020. Whether they will all go ahead remains uncertain. In 2011, Scottish and Southern Energy pulled out of one of the consortia, and in 2012 the German companies E.ON and RWE pulled out of another, citing the cost of the German nuclear phase-out as one reason (Reuters, 2012). That left EDF as the main player, with the UK government clearly keen to continue: in March 2012 France and the UK agreed on cooperative arrangements, including a (small) UK share in some of the equipment supply work.

It is still not certain that even this reduced programme will go ahead: the political leadership changes in France may affect Areva's and EDF's commitment. There were also attempts by opponents in the UK to block the UK government's support for the programme, for example by an appeal to the EU on the basis of state aid issues. In addition, the decision-making process has been challenged over the wider issue of whether these plants are actually needed; one report claimed that the government had been poorly advised about the level to which demand for electricity would rise (Bailey and Blair, 2012). Whether these interventions will succeed remains to be seen. They might be viewed as somewhat hamstrung by the at least grudging acceptance among the public of a new nuclear programme.

5.3 An anomalous response?

It is perhaps not so surprising, given the erosion of opposition and the acceptance of an expansion programme, that the UK public did not react strongly to Fukushima. As Professor Nick Pidgeon from Cardiff University has put it, the post-Fukushima situation in the UK was 'very different to the situation after Chernobyl where most people were against

DOI: 10.1057/9781137274335

nuclear energy and wanted it shut down' (Pidgeon, 2011). As a result, the outlook now in the UK seems to be at odds with what's happening in most of the Western EU.

This cannot be explained by party politics alone. The UK's right-of-centre Conservative–Liberal Democrat coalition government has embraced nuclear just as the preceding centre-left New Labour did. For comparison, it is interesting that it has been centre-right governments in Germany and Italy that have ended up with solid anti-nuclear policies. Of course, in both cases the spectacular policy U-turns were due to massive grassroots opposition to government nuclear policies (and/or leaders) post-Fukushima. A similar thing has occurred in France, with very pro-nuclear President Sarkozy ousted and the Socialist Party promising a partial nuclear phase-out.

Nothing like this has occurred in the UK. The only significant opposition was from the Liberal Democrats, who pledged to oppose nuclear power if elected but dropped this position once they became part of the coalition government, in return for an agreement that there should be no public subsidies. Indeed, despite previously being strongly opposed, Lib Dem Chris Huhne, who became Energy Secretary, took an increasingly pro-nuclear line. His replacement, Ed Davey, had a similar history of opposition converted to support. That has left the fledgling Green Party, with one MP, along with the eight Scottish National Party MPs and a few dissident members of the three main parties, as the parliamentary opposition to nuclear. The far left still opposes nuclear, while the far right has been very pro-nuclear: witness the strong support given by the British National Party leader, Nick Griffin MEP, in the European parliament.

Given this political background, it is perhaps not surprising that UK public reactions to Fukushima have also been relatively muted, at least according to the opinion polls.

An Ipsos MORI poll in June 2011 for the Nuclear Industry Association asked whether respondents agreed or disagreed that 'Britain needs a mix of energy sources to ensure a reliable supply of electricity, including nuclear power and renewables' – and 68% agreed. When asked more directly 'How favourable are you to the nuclear energy industry?', 28% said favourable, 24% unfavourable. When asked 'Do you support or oppose building new nuclear power stations to replace the existing fleet?', 36% supported, 28% opposed. So opponents were in a minority (Ipsos, 2011b).

However, to confuse matters, a poll produced for the British Science Association's Annual Festival in September 2011 found that opposition

DOI: 10.1057/9781137274335

was still in the majority, but had fallen. It said 37% of the population supported the use of nuclear for producing energy in the UK, but those 'very or fairly concerned' had reduced from 59% in 2005, to 54% in 2010, and to 47% in 2011, after Fukushima. It added that those 'not very or not at all concerned about nuclear' had risen from 38% in 2005, to 42% in 2010, and to 45% in 2011 (BSA, 2011).

No doubt the results depend on the questions used. For example, no poll seemed to ask whether a *large* new programme was wanted. An earlier Ipsos poll, in May 2011, found the UK split 48% for nuclear, 51% against, with 57% even saying the UK should 'stop future build'. Moreover, 74% disagreed with the idea of 'modernisation' of electricity production via nuclear, while a massive 80% felt that 'nuclear was not a viable long term option' (Ipsos, 2011a).

The nuclear industry, by contrast, had few doubts. After Fukushima, EDF insisted that plans to build a new generation of reactors in Britain would continue. Vincent de Rivaz, CEO of EDF Energy, said, 'While we understand the importance of adjusting the timetable to take into account the Nuclear Installations Inspectorate report [on the Japanese crisis], it is also equally important that establishing the framework for new nuclear should not be subject to undue delay. The events in Japan do not change the need for nuclear in Britain.' He spoke of 'determination to press ahead with our project, and the strong feeling that whilst we should learn any lessons we may need to from Japan, we should not delay our progress'.

The industry professed that it did not need subsidies, but it did want the market structure to be supportive. In 2010 the UK government proposed radical Electricity Market Reforms which some see as designed primarily to help sustain nuclear by creating a new 'Contracts for Difference' support system to replace the existing Renewables Obligation, and also by introducing a unilateral UK carbon price support system to buttress the EU Emissions Trading System (Mitchell, 2011). These changes may also help some of the larger renewables, but the nuclear industry is likely to benefit most; for example, its existing plants are likely to get a large windfall carbon credit bonus (Toke, 2010).

In the main, the media has taken the view that nuclear was the way ahead for the UK and that the safety issues were minimal. Perhaps inevitably, the dramatic and newsworthy aspects of the Fukushima disaster initially led some newspapers and news channels to adopt a sensationalist approach, but calming views concerning the implications were usually also presented, especially by the broadcast media.

DOI: 10.1057/9781137274335

Shortly after Fukushima, the BBC produced a number of radio and TV documentaries that, as we shall see, some critics felt were basically pro-nuclear, notably a *Horizon* documentary and an edition of its popular science show *Bang Goes the Theory*, both focusing on nuclear safety. The main message was that the science was clear: there had been relatively few deaths at Chernobyl, and none could be expected from radiation at Fukushima. The *Bang Goes the Theory* programme adopted the arguably rather limited view that what mattered was the hard, confirmed data on deaths, rather than speculation on possibly linked deaths. So, based on input from Professor Gerry Thomas from Imperial College, it claimed that there were just 122 deaths from Chernobyl and that 'zero' could be expected, from radiation, from Fukushima.

There were objections alleging bias, with several organisations submitting formal complaints (SGR, 2011; NCG, 2011). After protracted discussions with the complainants, the BBC subsequently formally admitted on its Complaints website that 'the figure 122 was presented as definitive whereas certainty is in fact lacking'. It concluded that 'the programme was misleading in that respect', although it added that this was 'not to a degree which might have amounted to bias in relation to the arguments about nuclear power' (BBC, 2012).

This final point is still disputed by the complainants. I will be returning later to look at some of the issues raised by this episode, since clearly the role of the media is important. However, possible media bias notwithstanding, the UK's stance on nuclear power might also be seen as surprising given that it has the best renewable resources in the EU. Independent studies have indicated that the UK could meet 50% of its electricity needs from renewables by 2020 (TPOES, 2008) and up to 88% by 2030 (WWF, 2011a), while one study has even suggested it could reach almost 100% by 2030 (CAT, 2010). It has been estimated that the offshore wind, wave and tidal resource alone could, if fully developed, supply six times current UK electricity requirements (PIRC, 2010).

So far though, by comparison with most other EU countries, the UK has been slow in developing its renewable resources. For example, while Germany has installed 29 GW of wind generation capacity, the UK, which has a much better wind regime, has managed only around 6 GW to date. The UK's renewable energy target for 2020 has been set much lower than those for many other EU countries, at 15% of total energy. By contrast, as noted earlier, Denmark is aiming for 30% and Austria

DOI: 10.1057/9781137274335

for 34%, both countries having similar climates to the UK. They differ, however, in not having nuclear programmes.

The difference in approach is further highlighted by the fact that, while the UK government has strongly pro-nuclear policies, Scotland's devolved government has maintained a strongly anti-nuclear line and has backed this up with a commitment to meeting 100% of Scotland's electricity needs from renewables by 2020, vastly beyond the target for the UK as a whole, and with no new nuclear (Scottish Government, 2011, 2012). Otherwise, the UK response to Fukushima, and its overall position on nuclear, seem to have been very different from the responses in much of the rest of Western Europe.

DOI: 10.1057/9781137274335

6
Reactions in the US and the Rest of the World

Abstract: *With the exception of the UK, energy policies and public opinion in most of Western Europe seem to have been significantly impacted by Fukushima. Reactions to Fukushima outside Europe were more mixed. As in parts of Asia, most governments stuck to their pro-nuclear policies, despite in some cases increasing levels of public opposition. This chapter looks at the US and then at the areas of the world not so far reviewed: the Middle East, South America, Africa and finally Russia. In all these areas, opposition rose, sometimes to very high levels, but policies mostly remained unchanged – with some notable exceptions in the Middle East.*

Keywords: African reactions; Middle East nuclear plans; Russian nuclear programme; South American reactions; US nuclear policy

Elliott, David. *Fukushima: Impacts and Implications.* Basingstoke: Palgrave Macmillan, 2013. DOI: 10.1057/9781137274335.

DOI: 10.1057/9781137274335

The tragic nuclear incident in Japan has introduced multiple uncertainties around new nuclear development in the United States.

David Crane, CEO of NRG Energy, on why they were
exiting a nuclear project in Texas (*ABJ*, 2011)

6.1 Reactions in the US

The US currently gets just under 20% of its electricity from its nuclear plants, most of which date from the 1970s. As we shall see, there had been plans for expansion, and some of these plans may still go ahead, although Fukushima may slow them, mainly because of extra costs and reduced investor confidence.

The public reaction to Fukushima in the US was somewhat similar to that in the UK. As in the UK, opposition to nuclear power had been high in the 1970s, and after Chernobyl, in 1986, it increased further, while support fell to 34%, according to a CBS News poll. Since then support has risen, as concerns about climate change and, in the US context more importantly, energy security have increased. In a survey by Bisconti Research for the Nuclear Energy Institute, carried out just before Fukushima, 71% favoured the use of nuclear. Bisconti said, 'it's clear that information about nuclear energy in the media is reaching substantial numbers of the public. And the public's view of nuclear power plant safety has transformed over the past decades. This research shows 67% of Americans viewing nuclear plants as safe, compared with 35% in 1984' (WNN, 2011f).

You could say this merely showed that PR works! But this survey was conducted before Fukushima, which led to dramatic shifts, according to some polls. For example, 64% moved to oppose nuclear in an ABC News/Washington Post poll, and, according to a CBS News poll, only 43% said they would approve the building of new facilities in the US. The usually authoritative Pew Research Centre said that support fell to 39%, with 52% opposed (PEW, 2011). The BBC GlobeScan poll came to similar conclusions: 58% in the US were opposed (44% to new plants, 14% to all plants), while 39% supported nuclear (BBC World Service, 2011). However, other polls gave different results.

A CNN/Opinion Research Corporation survey found that 57% of people questioned approved of the domestic use of nuclear energy, with 42% opposed. And a poll for the US Nuclear Energy Institute in November 2011 found that 62% still backed nuclear (NEI, 2012b).

DOI: 10.1057/9781137274335

In California, there were concerns about inland faults/quakes and the risk to coastal nuclear plants from tsunamis. But most US reactions were shaped by fear of long-distance contamination from Fukushima. There were reports of caesium contamination in drinking water and milk in some locations, although the authorities said it was below safety threshold levels (McMahon, 2011).

Some of these fears may have been misplaced, and some widely disseminated studies were criticised for being unduly alarmist and for poor handling of statistical data. Arguably, one case in point was a paper by Joseph Mangano and Janette Sherman in the peer-reviewed *International Journal of Health Services*. They noted 'an unusual rise in infant deaths in the north-western United States for the 10-week period following the arrival of the airborne radio-active plume from the meltdowns at the Fukushima plants in northern Japan'. Using data from US health officials' weekly reports, they suggested that there might be '13,983 total deaths and 822 infant deaths in excess of the expected', although they added that 'these preliminary data need to be followed up' (Mangano and Sherman, 2012).

There are uncertainties about the validity of their conclusions, and the study was widely attacked by the nuclear lobby, and in a blog linked to *Scientific American*, as unreliable and misleading (NEI Nuclear Notes, 2011; Moyar, 2011). This episode, and others like it, which can involve pronouncements from either side of the argument, indicate one of the problems for the wider public: when seeking informed views, whom do they trust? I will return to this issue later. But certainly the official view in the US remained that there were no significant problems from Fukushima or from US plants.

The US government initiated a review of nuclear plant safety after Fukushima, but has continued to support expansion, in part by offering loan guarantees to prospective private developers. However, so far these guarantees have not been too successful; for example, they were evidently not enough for Constellation Energy, which, as noted earlier, in 2010 pulled out of the proposed Calvert Cliffs-3 EPR reactor project in Maryland.

Following Fukushima, some other projects were abandoned or delayed. NRG Energy pulled out from investment in Units 3 and 4 of a project in South Texas. The company said, 'The tragic nuclear incident in Japan has introduced multiple uncertainties around new nuclear development in the United States which have had the effect of dramatically reducing the probability that STP 3&4 can be successfully developed in a timely fashion' (*ABJ*, 2011).

DOI: 10.1057/9781137274335

Subsequently, Progress Energy put back its plans for two Westinghouse AP1000 reactors in Levy County, Florida, by three years, by which time, it said, the cost will be $19–24 billion. It blamed the delay on 'lower-than-anticipated customer demand, the lingering economic slowdown, uncertainty regarding potential carbon regulation and current low natural gas prices' (WNN, 2012g).

However, some projects still look likely to go ahead. In early 2012 the US Nuclear Regulatory Commission (NRC) approved the first new nuclear reactors in the US for more than three decades. One member of the commission abstained from the decision, calling for assurances that the lessons of Fukushima would be fully applied. Southern Company has been given the go-ahead for two Westinghouse AP1000 reactors at the Vogtle nuclear site in Georgia. The reactors could begin operation in 2016 and 2017 (WNN, 2012h).

The NRC last approved construction of a nuclear plant in 1978, a year before a partial meltdown of the Three Mile Island nuclear plant in Pennsylvania. That accident was mostly contained, unlike Fukushima, but there were some small releases and permitted venting of radioactive material. There were assurances that there would be no health risks.

As in the UK, there are no specific targets for US nuclear expansion. In his State of the Union Address in 2011, Barack Obama said that nuclear power, clean coal and natural gas would all be needed, along with renewables, to meet a goal of 80% clean energy by 2035, but he did not specify the proportions of each.

However, in its provisional *Annual Energy Outlook 2011*, the US Department of Energy projected an increase in installed nuclear capacity of about 10 GW (10%) by 2035, of which 6.3 GW would be new capacity (five reactors), with the rest coming from up-rating. Given projected rises in energy demand and other supply options, the overall nuclear share would shrink from 20% to 17% (DOE, 2011).

So, although the government seems keen to expand nuclear, the expected expansion is relatively limited, and debate continues over its pros and cons (Ferguson and Settle, 2012). It could be that the poor economics of nuclear power will act as a disincentive to new developments in the US – and also in Canada: following Fukushima, Bruce Power decided not to pursue a new plant in the province of Alberta.

The head of Exelon, the largest US nuclear utility, told the American Nuclear Society's 2011 Utility Working Conference that the near-term prospects for US nuclear expansion 'will be miserably hard and extremely

DOI: 10.1057/9781137274335

challenged by economics. There is not currently a need for new base-load generation because of minimal load growth and excess generation capacity', and also an influx of shale gas.

A paper from the University of California, Berkeley concluded that, from the US perspective, 'it seems unlikely that there will be much of a renaissance' (Davis, 2011).

Opposition in the US may not be as extensive as in many other parts of the world, or even as much as in Canada, where the Ipsos poll put it at 63%, but it may be that Fukushima and the spread of renewables, along crucially with economic and waste disposal problems, mean that nuclear will be constrained and might decline (Jaczko, 2012).

6.2   Reactions in the rest of the world

Nuclear power has clearly been seen as a possible option for many developing countries, including, as we saw earlier, some in the Asia-Pacific region (Goldemberg, 2009). The developing world of course includes some countries that are much more developed than others, notably Brazil, China and India, which are sometimes labelled, along with Russia, as the 'BRIC' countries (although Russia is in effect undergoing rapid industrial redevelopment rather than initial development). I looked at China and India earlier. I look at Russia below. But first, in this final survey section, I look at reactions and policies in the Middle East, South and Central America (including Brazil) and Africa.

So far within these regions, only Argentina, Brazil, Mexico and South Africa have power-producing nuclear plants. Cuba abandoned its nuclear plans after the collapse of the USSR. Table 6.1 summarises the pre-Fukushima state of play for other countries in these regions, in terms of those that have indicted nuclear aspirations. Some of the commitments are speculative, although in some cases programmes are further advanced.

As noted earlier, there has been something of a nuclear push in the Middle East. For example, Saudi Arabia is considering a $100 billion programme to build 16 new plants by 2030. Abu Dhabi's first plant is due to open in 2017, with three more to follow. On the borders of the region, Turkey is also pressing ahead with nuclear plans. So, crucially, is Iran. As Table 6.1 illustrates, most other countries in the region have, at some point, expressed interest in nuclear power. Some have progressed quite far (Banks *et al.*, 2012). At one stage Iraq had nuclear ambitions, fuelling

TABLE 6.1　*Countries seeking to have nuclear power*

Middle East	Africa	South and Central America
Bahrain	Algeria	Bolivia
Egypt	Ghana	Chile
Israel	Kenya	Dominican Republic
Jordon	Libya	El Salvador
Kuwait	Morocco	Haiti
Oman	Namibia	Jamaica
Qatar	Nigeria	Peru
Saudi Arabia	Senegal	Uruguay
Syria	Sudan	Venezuela
Turkey	Tanzania	
UAE (Abu Dhabi)	Tunisia	
Yemen		

Source: Miller and Sagan (2009).

fears about the proliferation of nuclear weapons, which have deepened recently because of Iran's nuclear ambitions. Clearly the nuclear issue is a sensitive political topic in this region.

Fukushima seems to have produced critical public responses in most of these countries. For example, in the Ipsos post-Fukushima poll, 58% opposed nuclear power in Saudi Arabia and 71% in Turkey (Ipsos, 2011a). The BBC GlobeScan poll found that only 21% supported nuclear in Turkey, with 73% against (32% objecting to new plants and 41% to all plants; BBC World Service, 2011). Nevertheless, the plans in both countries presumably still stand.

However, Kuwait said it no longer wanted to go down the nuclear path (*Kuwait Times*, 2011). Bahrain subsequently also decided to halt its nuclear programme, with *Trade Arabia* reporting in February 2012 that, according to Dr Abdulhussain Mirza, Bahrain's Energy Minister, the government had reconsidered and abandoned the idea following Fukushima (Modern Power Systems, 2012). There were press reports after Fukushima that Qatar was also rethinking its nuclear ambitions, but that has not been confirmed. Indeed, recent industry indications have suggested continuing interest (Kanady, 2012). There has been continued opposition to the proposed nuclear plant in Egypt, and there were violent protests at the site in January 2012, but the programme is to continue (Nuclear News,

DOI: 10.1057/9781137274335

2012). The BBC GlobeScan poll found that only 21% of those asked were in favour of nuclear power in Egypt, while 66% were opposed.

Less was heard from the North African countries, embroiled as many were in the so-called Arab spring of political changes. As Table 6.1 illustrates, some had previously shown interest in nuclear power. Under Gaddafi, Libya developed some nuclear capabilities, and in March 2011 Algeria indicated that it was still thinking seriously about nuclear. Otherwise, nuclear power may have fallen off the agenda in much of the region. Certainly it has been suggested that local financial and institutional capacities are not sufficient to sustain a nuclear programme (Jewell, 2011).

In South America there were some strong reactions to nuclear after Fukushima. In Mexico 81% of those asked by Ipsos said they 'did not support nuclear power', 52% of them being strongly opposed. In Argentina 72% were opposed, and in Brazil 69% (Ipsos, 2011a). The BBC GlobeScan poll came up with even more dramatic results: 79% in Brazil were opposed to further or any nuclear plants (44% being opposed to new projects, 35% to any projects), and only 16% were in favour. In Mexico the figures were 82% opposed (39% against new projects, 43% against all). That survey also looked at Ecuador, where 65% were against and just 6% for; Chile, with 81% against, 3% for; Peru, with 53% against, 15% for; and Panama, with 71% against, 11% for (BBC World Service, 2011).

In sub-Saharan Africa, South Africa already gets 6% of its electricity from nuclear power, and had been planning to expand its capacity. But the recession led it to abandon the programme, which included its novel 'pebble-bed' mini-reactor. However, this halt may be temporary: South Africa appears keen to continue with its nuclear programme. But it also sees renewables as making an even larger contribution. In the Ipsos poll following Fukushima, 60% of South Africans were opposed to nuclear power (Ipsos, 2011a).

In Kenya, the government seems keen to focus very heavily on nuclear power. It has plans for a $1 billion programme which, if it went ahead, would supply most of the country's power. That is some way off, and concerns have been aired about the cost. In the BBC GlobeScan poll 39% of those asked opposed nuclear power outright, with 15% opposing just new plants, and only 29% supporting nuclear. The poll found that support for nuclear was a little higher in Nigeria (41%) and also in Ghana (35%). However, opposition was also strong, at 33% in Ghana and 48% in Nigeria, with roughly equal numbers opposing either new nuclear or any nuclear (BBC World Service, 2011).

DOI: 10.1057/9781137274335

Overall, with some exceptions, it seems that, after Fukushima, nuclear is not particularly popular in Africa, and is even less popular in the Middle East, while in South America it is widely and quite strongly opposed. However, that may not translate through to government policies in these regions. For example, most (but not all) nuclear programmes are still continuing in the Middle East, where some are quite well developed, and although Brazil announced in 2012 that it was delaying the start of its proposed new reactor, no major policy changes seem to have emerged in either Africa or South America.

The situation in Russia is somewhat similar, with strong public opposition but also strong government support. But Russia starts from a very different technological base. It already has a large and growing national nuclear programme, as well as a major role in exporting nuclear technology worldwide, which it is keen to expand. Vladimir Putin has said that nuclear is the only alternative to traditional energy sources. He commented, 'You couldn't transfer large electric power stations to wind energy, however much you wanted to. In the next few decades, it will be impossible.'

Putin said that energy consumption patterns will undergo only minor changes in future. Nuclear is, he said, the only 'real and powerful alternative' to oil and gas, and he called other approaches to meeting future energy demand 'claptrap' (Cook & Elliott, 2012).

At present Russia has just 9 MW of wind capacity in place, and a target of getting about 4.5% of its electricity from renewables by 2020. This is very low compared with its large renewable potential. As we shall see, that potential could run into the hundreds of gigawatts just from wind. By contrast, Russia currently has nearly 22 GW of nuclear capacity, and is pressing ahead with a major nuclear expansion programme, hoping to nearly double output by 2020. However, it may have to contend with popular opposition. After Fukushima, the BBC GlobeScan poll found that opposition to nuclear had risen from 61% (in 2005) to 80%, with 43% opposing nuclear outright and 37% opposing new nuclear plants, and just 9% being in favour.

DOI: 10.1057/9781137274335

7
Analysis: Political, Economic and Technological Issues

Abstract: *While the immediate responses to a major accident like Fukushima may reflect concerns about personal safety and health, reactions to Fukushima, and to nuclear power in general, are also likely to be shaped by a variety of other factors, including views on technology, political orientation and economic assessments. Drawing on the review of reactions in previous chapters, this chapter explores possible explanations for the pattern of public and governmental reactions that emerged, and looks at what might happen next, asking whether new approaches to energy will emerge.*

Keywords: energy policy choices; nuclear economics; political factors; social factors

Elliott, David. *Fukushima: Impacts and Implications.* Basingstoke: Palgrave Macmillan, 2013. DOI: 10.1057/9781137274335.

DOI: 10.1057/9781137274335

> There have so far been no radiation-induced deaths, nor are there likely to be any in the future. How can this relatively benign incident create such a degree of fear that it is dominating discussion of nuclear power's future?
>
> Steve Kidd, World Nuclear Association (Kidd, 2012a)

7.1 Reactions to Fukushima – and to nuclear power

The previous chapters have reviewed reactions to Fukushima around the world. It has to be said that, even at the time of writing, just over one year after the accident, there are still major problems at the reactor site. The disaster may not be over. For example, there seem to have been continuing problems with leakage of contaminated cooling water (Dvorak, 2012) and fears that, despite 9 tonnes of water an hour being pumped in, cooling water levels inside some reactors may be too low, exposing core components, which might reheat and re-melt and possibly release more radiation. An internal remote inspection in March 2012 found that radiation levels were much higher than expected, up to 60 Sv, and there were continuing concerns about the condition of the spent fuel rods: some are still in a waste storage pool on top of the crippled Reactor 4 building, 100 feet above the ground. Kazuhiko Kudo, a professor of nuclear engineering at Kyushu University, was quoted by the *New York Times* as saying, 'The plant is still in a precarious state. Unfortunately, all we can do is to keep pumping water inside the reactors, and hope we don't have another big earthquake' (Tabuchi, 2012). In addition, problems remain outside the reactor buildings, with fragments of nuclear fuel rods or pellets and other highly active materials scattered around the site (ENE News, 2012). Moreover, the contamination of land and sea could still be an issue for Japan and for neighbouring countries.

Although the story might not be finished, with over a year elapsed (at the time of writing) it is reasonable to review, as in the following sections, the implications of the accident and the reactions to it. As we have seen, some countries have significant nuclear programmes, although most do not; in either case, opposition is common, and it increased after Fukushima. Can patterns in the responses be identified?

Poll data provides snapshots of people's opinions, with varying degrees of reliability, but it usually does not indicate their reasons for holding these opinions. After an accident like Fukushima some of the responses among people in the immediate area may be due to straight-forward

DOI: 10.1057/9781137274335

personal fear of health hazards or, at a lower level, concerns about disruption, risks to employment, house prices and so on. But for others, more geographically remote, more general factors may be involved, reflecting views about nuclear power developed over time. Not all of these factors are negative. For example, the adoption of nuclear power has sometimes been portrayed as representing a scientific and technological advance, and as part of a progressive economic modernisation programme. More recently it has been portrayed as a low-carbon option, although there are disputes about whether it is, or will remain, low-carbon, given the energy needed to produce its fuel (Sovacool, 2008; AIE/AEA, 2011).

Wider strategic issues such as these may inform the debate, but there are also other, perhaps more immediate concerns. In what follows, I look at some key factors that may influence support for and opposition to nuclear power in general and reactions to Fukushima in particular.

Attitudes to nuclear power have been much studied in the past, with the focus usually on the growth of opposition movements as a political or sociological phenomenon (Surrey and Huggett, 1976; Kitschelt, 1986). Fukushima seems to have initiated a new phase in this social movement. However, the prevalence and effectiveness of opposition movements varies around the world, depending on a range of local factors and other, wider concerns. For example, following Fukushima, a US commentator insisted that '[t]he U.S. dependence on foreign oil from the Middle East is a greater threat to America than the use and expansion of nuclear power. If the price of gasoline at the pump heads closer to $5 a gallon in coming weeks, fears over nuclear power will dissipate' (Botta, 2011).

Views like this may explain why many in the US are continuing to support nuclear power, even if this perhaps rather simplistic analysis does not make much sense in energy terms. Nuclear electricity would mostly replace electricity generated using coal and possibly gas; very little oil is used for this purpose. But, to be fair, maybe Botta was thinking that nuclear electricity might help power battery electric vehicles?

It does seems odd that the vast potential for renewables in the US is often not seen in the same light, especially since, as well as electricity for battery-powered vehicles, they could produce liquid or gaseous biofuels for use in vehicles.

A similar techno-economic viewpoint was revealed by the head of Italy's Nuclear Safety Agency, Umberto Veronesi, who, after Italy's referendum opposing new nuclear, said, 'My fear is that Italy will finish as a tourist appendix to the advanced world.'

DOI: 10.1057/9781137274335

7.2   Technological choices

What seem to be at work here are a basic prejudice against renewable energy technology and an enthusiasm for what is seen as 'high-tech' centralised nuclear. This is very apparent in Putin's remarks quoted in the previous chapter – and this in a country which has vast wind and hydro resources. For example, it has been estimated that there could be 350 GW of wind capacity in the high, cold, windswept steppes of north-west Siberia and northern Russia, which could generate 1,100 TWh p.a., more than Russia's total 870 TWh electricity production in 2009. The power would have to be brought back to the west of the country, where it is most needed, via high-voltage direct current supergrids, but this is just the sort of 'high-tech' project you would think would appeal to Putin.

Some of the former Soviet Bloc countries in the Near East also have very large renewable potentials, notably Kazakhstan, which, it is estimated, could have 210 GW of wind capacity, capable of generating 550 TWh p.a. It also has major hydro potential. For the moment, however, Kazakhstan is focused almost exclusively on its oil, gas and uranium resources, and it has talked of developing nuclear power. It is a major exporter of uranium, and its fossil fuel resources will obviously put it in the front line of Western interest, for as long as these reserves last. But, given that this will not be for ever, it would seem to make sense to start to exploit the huge renewable resources it has now.

Energy and technology issues are also an important element in the support for nuclear in the Middle East, which has much to do with concerns that demand for energy is increasing as the region becomes more affluent, and especially given the increasing use of air-conditioning. Desalination plants also need power. Climate change will make matters worse. A recent study suggested that, 'on the current trajectory, Saudi Arabia's domestic energy consumption could limit its exports of oil within a decade' (Lahn and Stephens, 2011).

Nuclear power may seem an odd choice in a region blessed with consistent and strong sunshine and also with plenty of room in desert areas for solar energy collectors of various types. A study by Bloomberg New Energy Finance claimed that the falling costs of PV technology meant that solar energy was already a more economically attractive option for domestic electric power generation in the Gulf Cooperation Council region than oil-fired electricity production. Part of the attraction is that

DOI: 10.1057/9781137274335

it would free up oil to be sold at world market prices, more than offsetting the cost of solar (Bloomberg, 2011).

The solar idea has already caught on, with a mix of solar technologies being used. For example, Dubai Electricity and Water Authority is to build a 1,000 MW solar park project southeast of Dubai city, using both PV and concentrated solar power (CSP).

Many other Gulf states are also looking at solar power as a way to diversify from oil. Overall, the programmes have been relatively small scale so far. For example, the UAE is planning to get 7% of its power from renewables by 2020. It has a 100 MW PV array. But several CSP arrays are being planned, for example, in Jordon, Morocco and the UAE. Egypt already has a 150 MW hybrid solar–gas project, and more are planned. The Egyptian National Plan for 2012 includes a 100 MW CSP plant in south Egypt, and the National Plan for 2018–2022 has 2,550 MW of CSP. Egypt aims to get 20% of its power from renewables by 2020.

Although most of the CSP plants at present have gas-fired backup, molten-salt heat stores can be used to capture some of the daytime solar heat to run the generators overnight, so offering 24/7 power. Longer term there are prospects for exporting some of the power to the EU, thus directly replacing oil revenue.

It is conceivable that nuclear could provide the same benefits, even if its fuel has to be imported. But, in addition to the security problems of operating in the Middle East, there is more than a hint of suspicion that one of nuclear's attractions is the ability it confers to produce nuclear weapons, should that be deemed necessary at some point.

The Saudis have already indicated that they might be forced to do this if Iran develops nuclear weapons. To date, all the known nuclear weapons in the world have been produced by states (Israel apart) with civil nuclear technology; the technologies are inevitably intertwined, often making it hard to detect which option is being emphasised (see Section A.2 in the appendix).

Even discounting that issue, there could be practical problems. Steve Kidd from the London-based World Nuclear Association, writing in Nuclear Engineering International's newsletter, has suggested that nuclear 'may not for some years be suitable for many countries that do not have the developed institutional framework to cope with it. Such a description must apply to most of the Middle Eastern countries currently looking at nuclear power' (Kidd, 2010). He continues, 'Maybe it would be far better for them to specialise in developing solar power and

DOI: 10.1057/9781137274335

other renewable solutions, combined with developing their power grids to cope with diffuse and sometimes intermittent technologies.' It is hard not to agree.

Similar arguments may apply elsewhere in the developing world, in Africa and Latin America. Even in the newly industrialised East it is hard to see why nuclear is to be preferred to renewables, given that most of Asia has plentiful renewable resources.

As we have seen, China is embarking on a major renewables development programme, with nuclear playing only a relatively minor role. Chen Mingde, vice chair of the National Development and Reform Commission, in comments quoted by the *China Daily* newspaper in 2010, claimed that 'nuclear power cannot save us because the world's supply of uranium and other radioactive minerals needed to generate nuclear power [is] very limited'. He saw the expansion of China's nuclear power capacity as a 'transitional replacement' for its heavy reliance on coal and oil, with the future being in more efficient use of fossil fuels and expanded use of renewables such as wind, solar and hydro.

India is expanding its renewables programme in parallel with its nuclear programme. It is often said that India's nuclear programme has been rather slow to get moving. But it is now among the leaders in wind power. By contrast, despite its huge renewable potential, Australia is still just starting out. Until recently it has focused mainly on using fossil fuels and exporting uranium.

Despite dragging its feet on climate change policy, the US is already a leader in renewable energy, second only to China. It has set a target of supplying 15% of US power from renewables by 2025 and clearly sees leadership and innovation in this field as key to its future economic success. As Barack Obama put it at the time of his election, '[T]he country that harnesses the power of the clean, renewable energy will lead the 21st century.'

Technological issues and priorities clearly play their part in shaping policies on, and responses to, energy in general and nuclear power in particular, and they are to some extent linked to geography: location defines some of the key strategic factors related to access to energy resources. It may also shape specific reactions to nuclear power. For example, Greenpeace Germany made the telling point that, unlike Japan, Germany 'could not pump radioactive water into the sea', at least not from the majority of its plants, which are far away from its northern coastline.

DOI: 10.1057/9781137274335

7.3   Political orientations

Technology and location are not the only issues. National politics also clearly play a role (Sovacool and Valentine, 2012). For example, Denmark and the Netherlands are neighbours with similar geographical situations and histories, so you might expect similar responses to nuclear. But there have been clear divergences, due in part, it seems, to political differences. Maintaining its strong anti-nuclear position, the new centre-left government in Denmark is backing a target of obtaining 50% of its electricity from renewables by 2020, and is aiming to reach 'zero carbon' by 2050. By contrast, in the Netherlands, where both nuclear and renewables had been supported in the past, a new centre-right government wanted to remove subsidies for renewables, but backed a new nuclear plant. So politics can have a significant impact. This becomes especially clear when there are changes in political direction, as happened recently in Spain.

Spain's recent shift to the right may undo the significant progress it has made in the renewables field. Spain had become a world leader in wind and solar PV, but in response to the recession and the euro/credit crisis, in January 2011, as an emergency measure, the new government cut all subsidies for renewables. Moreover, despite Spain's strong anti-nuclear movement, its long-standing nuclear phase-out programme, based on a ban on new plants and early closure of some older plants, might be abandoned. An agreement has been reached to allow the lifetime of some older plants to be extended, in part because running old plants longer is a low-cost option, and economic pressures are clearly high in Spain.

The role of politics can be overstated. Although, arguably, renewables have done better under left-of-centre governments, Denmark has pressed ahead with renewables and opposed nuclear under a range of left and right governments (Sorensen, 2011).

Moreover, as we have seen in Germany and Italy, there are now examples of right-leaning governments opposing nuclear and supporting renewables. Strongly anti-nuclear Austria has mostly had right-leaning governments, and both the far right and the left in France are now anti-nuclear. In the UK in the 1980s, Labour opposed nuclear, as did the right-leaning Conservative government. But both later went pro-nuclear.

Similar patterns can be seen elsewhere. Although US liberals have usually been critical of nuclear, both Democrat and Republican administrations have tended to issue pro-nuclear statements at irregular

DOI: 10.1057/9781137274335

intervals. In 2008, while running for office, Barack Obama made some commitments to developing 'green energy', but in the event that concept seems to have been elastic enough to include nuclear. However, during the current election year, presumably as a result of the adverse publicity surrounding Fukushima, none of the Democrat or Republican contenders for the presidency has shown much interest in the nuclear issue.

While it seems true that, in general, pro-nuclear views tend to be associated with right-wing or authoritarian positions, this is clearly not a simple ideological or party-based polarity. Most obviously, nuclear power continued to be supported strongly by governments in Russia both before and after the fall of communism, although some might describe both regimes as authoritarian. Cynics might add that, in the West, in a perhaps perverse reversal of the Russian experience, it was sometimes said by the hard left that nuclear power was dangerous under capitalism but would be safe under socialism!

7.4   Other patterns of response

While politics and political orientation clearly can be important in shaping nuclear policies and reactions to them, making simple links between ideology and approaches to nuclear power may not always be helpful. There are likely to be other factors at work, too. As seemed to be the case during the previous phase of nuclear opposition, following Chernobyl, there may be no single determining factor shaping responses (Koopmans and Duyvendak, 1995). Can we detect more general patterns in public responses and attribute them to specific causes or influences?

One consistent result from poll data is that women are more likely to oppose nuclear than men. For example, a poll carried out for the British Science Association in 2011 found that 53% of men were in favour of nuclear power, compared with only 21% of women, while 39% of women were opposed, but only 23% of men (BSA, 2011). Young people are also more likely to be opposed, as was found in a DECC Youth Panel consultation exercise (DECC, 2010).

However, in general there are problems with interpreting poll data, beyond these basic factors. Studies at the University of Cardiff and elsewhere (Pidgeon *et al.*, 2008a, 2008b) have shown that responses to general questions about energy-related policies may be very different from responses concerning specific events and decisions. So while people may

DOI: 10.1057/9781137274335

say they are happy with nuclear in general, when accidents occur they may respond differently.

Views are also likely to vary if the questions put to respondents ask about *relative* levels of support for nuclear compared with other options. For example, while in a 2008 Cardiff University/Ipsos MORI poll 71% believed that the benefits of nuclear power outweighed the risks, most respondents wanted more investment in renewables, with 71% saying that promoting solar and wind power was a better way of tackling climate change than nuclear power. The research leader, Professor Nick Pidgeon from Cardiff University, commented, 'In terms of developing a low carbon energy economy for Britain, renewables are clearly favoured whilst nuclear power remains unpopular but may be accepted alongside the development of a range of other energy sources' (Pidgeon, 2008).

A 2007 meta-study of 23 UK polls covering the period 2004–2007 by the Parliamentary Office for Science and Technology came to similar conclusions: 'When considered on its own, people were negative about nuclear energy, but became more positive when it was considered in combination with other technologies, such as renewables or energy efficiency measure.' It added, 'There is some evidence that support for nuclear power has increased over recent years, perhaps due to arguments relating to energy security and its reframing in terms of climate change mitigation.' As an example it noted that 'some respondents indicated that they might support nuclear power if it would help mitigate climate change. However, renewable sources of energy were seen to be a better means of doing this' (POST, 2007).

While views about technology choice may drive some people, with some clearly seeing renewables as the 'softer', less aggressive option, there may also be more complex contextual factors at work, related to the social and political environment. For example, one (perhaps speculative) explanation for the strength of the anti-nuclear movement in Germany goes back to its founding in the 1970s. At that time there were strong and arguably repressive prohibitions against membership of a range of proscribed political organisations, left and right. This meant that it was hard for members of such groups to get professional work in state agencies, schools and so on. This situation might partly explain the growth of the anti-nuclear and Green movements: they were radical 'single-issue' campaigns which were respectable, or at least not illegal. They offered a new context for radical opposition.

Something similar arguably happened in the Eastern bloc. For example, a strong anti-nuclear movement emerged in Lithuania during the

DOI: 10.1057/9781137274335

last years of what is now called the Russian occupation. The ostensible target was the Ignalina nuclear plant built by Russia near the Russian border, with much of the power allegedly going to Russia. While the rhetoric of the campaign focused on the problems of nuclear power, there was also an undertow of resentment at what was portrayed as Russian imperialism. Given the power of the state security services, direct opposition to Soviet rule was politically and personally danger-ous, and the anti-nuclear movement might be seen as something of a proxy or surrogate for resistance to Russian dominance. Interestingly, after Lithuania became independent in 1990, much less was heard from the once-powerful anti-nuclear movement. Indeed, some of its leaders went into the new national government, which then began to portray the Ignalina plant as a key national asset (Elliott and Cook, 2004).

These examples reinforce the view that it is hard to use simple political affiliations as a guide to reactions to nuclear power: things can be more complex that that. Certainly, as we have seen, changed political and economic circumstances may lead to differing responses, with no clear political pattern emerging. It may be that radical oppositional move-ments emerge, dialectically, in repressive environments, but there may also be other drivers and orientations. For example, it might be argued that what we sometimes see, as in some of the cases where right-wing governments have opposed nuclear, is simple political opportunism, although, in the French context at least, it could also be claimed to be a rational response to the allegedly failing economics of the French nuclear industry.

With that in mind I now look briefly at economic issues, which may offer a more solid basis for understanding responses to nuclear than the more general ideological, social and political orientations I have been looking at, especially given the difficultly, in the modern world, of defining ideological orientations, when terms such as 'left' and 'right' are sometimes less useful.

7.5   Economic issues

While reactions to nuclear power are sometimes based on safety or security issues, especially after events like Fukushima, or after 9/11 in the US (when there was much debate about the threat of terrorist attacks),

DOI: 10.1057/9781137274335

economic issues are often seen as having more underlying significance, given the high cost of building nuclear plants.

However, there are links between the safety and economic issues: historically, nuclear costs have been strongly influenced by safety concerns, and this seems likely to be the case for Fukushima (Cooper, 2012). Certainly there will be very high clean-up costs. Nevertheless, there have been claims that 'one-off' disaster-related costs, even large ones, could be borne over time, because the income from running nuclear plants can be very high. The cost of accidents, it is suggested, spread over time, adds just a small amount to bills. For example, Japan's Institute of Energy Economics said that even with compensation of up to $130 billion for the dislocation caused by Fukushima, the cost of nuclear generation, currently put at around $0.09/kWh, only increases to $0.11/kWh, whereas the cost of electricity from fossil fuels over the past five years has averaged $0.13/kWh, and from renewables (mostly geothermal) $0.12–0.11/kWh (WNN, 2011g).

This comparison may be suspect. There is more to renewables than just geothermal; for example, wind is much cheaper. Moreover, as we shall see, $130 billion is a low estimate for the dislocation and clean-up costs. This analysis is also based on what might be seen as optimistic estimates of the likely future costs of nuclear, especially given the need for extra safety measures. Crucially, it also ignores the wider cost of Fukushima and likely knock-on effects on insurance costs.

The insurance issue has global implications. Nuclear projects are hard to insure and, although plant operators have to provide some cover, governments have to step in to cover the full potential damage liabilities. Following Fukushima, an Associated Press/Washington Post report noted that a worst-case nuclear accident in Germany could cost €7.6 trillion, while the mandatory reactor insurance was only €2.5 billion. In Switzerland, the obligatory insurance was, it said, being raised from 1 to 1.8 billion Swiss francs ($2 billion), but a government agency estimated that a Chernobyl-style disaster might cost over 4 trillion francs, around eight times Switzerland's annual gross national product. In the US, insurance for nuclear operators is capped at $375 million by law, with further claims funded by the utilities up to $12.6 billion. France requires insurance of only €91 million from plant operators, with the government guaranteeing liabilities up to €228 million. The UK government may raise its liability requirement to £1 billion per event, taxpayers covering the rest.

These sums have to be compared with an estimate by the Japan Centre for Economic Research, which said that the full cost of Fukushima, including compensation for those displaced, could be up to $250 billion over the next 10 years (News on Japan, 2011). For comparison, although it has sometimes been (perhaps unkindly) suggested that its claims were inflated to attract aid, Belarus has estimated its economic losses due to the cumulative health and social impacts of Chernobyl over 30 years at $235 billion, and it has been reported that 5–7% of government spending in Ukraine still goes to Chernobyl-related benefit programmes (Green Facts, 2012).

Insurance liabilities and damage costs are, of course, only one aspect of nuclear economics. The central issue is the high initial capital cost and its relation to generation costs. Unfortunately, energy economics can be complex and contentious, and this is not the place to explore them in any detail or, for example, to investigate why lower costs are often quoted for nuclear projects in the East than in the West, or why some argue that, over the years, nuclear investment costs globally have increased and could well continue to do so (Cooper, 2009, 2012).

While there are often strongly divergent views influencing price comparison exercises, several studies have suggested that at present, depending on the location, there is not a lot to choose between nuclear and on-land wind. Indeed, in some locations wind is clearly the cheapest energy supply option on the grid (Sourcewatch, 2009).

Cost comparisons depend on assumptions made about interest rates, subsidy levels and technological progress, which become debatable when we look to the future, so there are a range of views, sometimes reflecting political orientations. For example, the UK's right-leaning Adam Smith Institute and the CIVITAS think-tank have both produced studies claiming that support for wind power was misguided and calling for more emphasis on nuclear (ASI, 2011; CIVITAS, 2012). This has also been a common theme in the right-wing press in the UK.

Certainly some commentators see nuclear as the best option longer term. Even so, in a report for the UK Department of Energy and Climate Change, consultants Mott Macdonald said that, for the UK, 'onshore wind is the least cost zero carbon option in the near to medium term, with a central cost estimate of £94/MWh, some £5/MWh less than nuclear on a FOAK [first-of-a-kind] basis' (Motts, 2011a).

Looking ahead, though, they said that 'while offshore is projected to see a large reduction in costs, compared with onshore wind, it will still face much higher costs at £110–125/MWh for projects commissioned from 2020', while nuclear could get cheaper. But they admit to taking a very bullish line on nuclear costs, and arguably are quite pessimistic about wind costs. For comparison, the European Wind Energy Association has estimated that, averaged across Europe, nuclear will cost €102/MWh in 2020, while onshore wind energy will cost €58/MWh and offshore wind €75/MWh (EWEA, 2012).

Other renewable options may also begin to challenge nuclear. Wave and tidal current technologies are developing rapidly around the world, with the UK in the lead at present. The UK's Carbon Trust has commented that 'with targeted innovation energy generation costs for both wave and tidal stream technologies could reduce to an average of 15 p/kWh by 2025, equivalent to today's cost of offshore wind energy'. It added that, with continued targeted innovation, 'the UK's best marine energy sites could generate electricity at costs comparable with nuclear and onshore wind', perhaps as soon as 2025 (Carbon Trust, 2011).

Solar PV is more expensive at present, but it has been predicted that it will reach 'grid parity' within a few years in some locations, and it is widely expected to become a major energy source globally. Mott Macdonald saw it becoming one of the most economic renewable options for the UK by 2040, and the UK is far from being the best location for PV (Motts, 2011b).

Biomass-based technologies are also making their mark, although there are land-use and biodiversity constraints on how much they can contribute (e.g. to providing vehicle fuel) and also issues about importing biomass for combustion plants. But biogas production from agricultural and municipal waste looks very promising, for heat as well as electricity production. There are a range of other heat options, including solar. At present, solar heat capacity is at nearly the same level globally in energy terms as wind power, and there are many interesting heat storage projects, including solar-fed community-scale district heating schemes, some with inter-seasonal heat stores (Elliott, 2011a).

Overall, it seems likely that renewables will continue to develop and to become more economic, while only the most bullish of nuclear supporters claim that nuclear will reduce in price dramatically. This view even seems to be accepted by Exxon Mobil, not the most obvious supporter of renewables, which sees the cost of electricity from on-land wind as becoming lower in the US than the cost of electricity from nuclear or fossil fuels with carbon capture and sequestration by 2030 (Exxon, 2012).

DOI: 10.1057/9781137274335

In addition to the climate and energy benefits, one of the attractions of renewables is that investment in them can create jobs rapidly. Many of the options are modular and involve relatively small projects, distributed widely across countries, which can be completed relatively quickly, in months rather than several years as with large centralised nuclear projects. Over 2 million people were working in the renewable energy field globally by 2008. A United Nations Environment Programme study suggested that 8 million jobs would be created globally in wind and solar alone by around 2020 (UNEP, 2008).

Investment in nuclear would also create jobs and, it is argued, deliver economic benefits, including benefits from exporting technology around the world. However, as France has found, there are uncertainties about how secure these gains might be.

Changing attitudes and policies in some countries, post-Fukushima, clearly make nuclear an uncertain investment option.

The UK was also impacted economically by Fukushima. It handled Japan's spent fuel, extracting plutonium from it at the reprocessing plant at Sellafield and then converting some of it into MOX (mixed plutonium and uranium oxide fuel) for use in some of Japan's reactors. Some MOX was in use at Fukushima (about 95 tonnes), although it was not supplied by the UK. After Fukushima, given that Japan probably will not be needing any more MOX from any source, the £1.2 billion Sellafield MOX fabrication plant was closed. It had in any case not worked well.

Fukushima also led to massive economic dislocation in Japan. Quite apart from the clean-up costs, the extended nuclear shutdown led to loss of revenue for the power companies, and a large extra bill for importing stopgap fuel to run conventional plants, put at $55 billion in 2011. This tipped Japan's trade balance into the red for 2011. Bloomberg reported that six Japanese power companies had losses of $6 billion due to increased fossil fuel costs and idled nuclear capacity (WNN, 2012c).

Renewables are unlikely to face major industry-wide shutdown crises like this. They are seen as offering prospects for sustainable employment, with, as they expand, job growth which would more than counterbalance the job losses from abandoning nuclear and phasing out fossil fuel. There would also be financial benefits from the reduction in the need to import fuel, which would counterbalance the cost of making the transition. An EU-wide study said that achieving the EU's 2020 target of a 20% energy contribution from renewables would lead to a net increase in GDP of about 0.24% (EC, 2009).

DOI: 10.1057/9781137274335

At the global level, a 2011 report by the Intergovernmental Panel on Climate Change suggested that renewables could supply up to 77% of global energy by 2050, with the right enabling public policies. The IPCC added that, although some renewables were already competitive, 'if environmental impacts such as emissions of pollutants and greenhouse gases were monetized and included in energy prices, more renewable energy technologies may become economically attractive' (IPCC, 2011).

7.6   Energy choices: the contested future

The increasingly positive technical, economic and strategic case for renewables may be beginning to win through, to the detriment of nuclear power. Even so, it is still a slow process. Fukushima might provide an impetus for a more rapid rethink. Indeed, tragically, it could be that major events like this will succeed in changing policy and viewpoints more rapidly than rational energy and strategic analysis.

However, this is not an automatic process. As we have seen, there have been some changes in policy following Fukushima, in some cases dramatic ones. But, despite growing opposition, representing a majority in most countries, many governments around the world still back nuclear power. Moreover, some commentators have suggested that 'nuclear is too big to fail' and that the industry will regroup, in Japan and elsewhere (Hoedt, 2012). Certainly Asia is seen as a major area for growth (Kidd, 2011).

A 'Nuclear Energy Market Outlook to 2020' report in 2012 said that 'construction of new nuclear power reactors, planned and proposed reactors, decommissioning projects and existing fuel cycle plant will open opportunities to nuclear suppliers and service providers globally'. It included Japan in this prognosis. It added, 'The nuclear power market in Asia-Pacific is expected to be the next global nuclear powerhouse', with the market expected to increase 'at a compound annual growth rate of 8.9% over the period 2011–2020' (Nuclear Intelligence Reports, 2012).

This may be speculative, but it does remind us that the nuclear story is far from over. Globally, R&D spending on nuclear dwarfs that on renewables and energy efficiency (Table 7.1). The nuclear lobby no longer dismisses renewables out of hand, as it sometimes did in the past, but still often claims that talk of them being able to meet all or

DOI: 10.1057/9781137274335

TABLE 7.1 *Government-funded energy subsidies in International Energy Agency countries (2007 US$ million)*

Technology	1974–2007		1998–2007	
	Cumulative total	% share	Cumulative total	% share
Energy efficiency	38,422	8.9	14,983	14.2
Fossil fuels	55,072	12.8	11,114	10.6
Renewable energy	37,333	8.7	10,709	10.2
Nuclear fission/fusion	236,528	54.8	43,667	41.5
Hydrogen & fuel cells	2,824	0.7	2,824	2.7
Transmission and storage	15,717	3.6	5,388	5.1
Other	45,204	10.5	16.599	15.8
Total	430,855	100	105,194	100
Nuclear's share of total subsidies, by country (%)				
Canada	39.0	28.8		
France	81.4	72.5		
Germany	67.0	41.0		
Japan	72.7	67.2		
Sweden	15.2	6.7		
UK	69.0	32.7		
US	38.1	13.2		

Source: Sovacool (2011).

most of our global energy needs is misleading. There is no doubt that would be challenging, especially given the imbalance in funding. As Table 7.1 shows, renewables remain the poor relation. Nuclear gets over four times more in subsidies than renewables, and three times more than energy efficiency.

It is sometimes claimed that nuclear research is more expensive but will yield more energy ultimately. However, most of the funding imbalance seems to be more to do with the continuing power of the nuclear lobby. Whether it will continue to be able to dominate R&D allocations (as in Japan and France, for example) remains to be seen.

As these figures illustrate, energy priorities around the world clearly differ. As we have seen, some governments see nuclear as key to maintaining their energy security, especially in countries where energy independence is considered to be important strategically (Ipsos, 2012).

DOI: 10.1057/9781137274335

This is not to say that the nuclear industry is not vulnerable. As Steve Kidd has put it, since the Fukushima accident,

> public and political acceptance of nuclear power has taken something of a knock in certain countries, resulting in the revival of phase-out policies. Germany is the most notable example of this but Switzerland and Belgium have also followed, albeit more gently, on a similar tack. Even in countries where nuclear power is still being endorsed as a useful contributor to a clean energy future, statements in support of nuclear power have been something less than strong and positive endorsements. And if this is the case, political choices in favour of nuclear and decisions made by private companies to invest in new nuclear stations are likely to get deflected by the slightest problem, and other less worthy energy options may indeed be pursued. (Kidd, 2012a)

His view remains that much of this reaction is mistaken. Fukushima was 'the worst nuclear disaster in 25 years, with radiation releases and contamination in some communities. And yet there have so far been no radiation-induced deaths, nor are there likely to be any in the future. How can this relatively benign incident create such a degree of fear that it is dominating discussion of nuclear power's future?'

He says, by way of explanation, that it is due to mistaken beliefs about the impacts of radiation. And more specifically he says, 'There is undoubtedly a huge economic impact of moving people from their homes and jobs in order to protect them, but the reason for this is the essentially unwarranted fear of radiation. This in itself can cause many illnesses, providing much wider implications a long way from the scene of the accident.'

He thus seems to put much emphasis on psychological impacts. He admits that, in reacting to concerns from the public, the nuclear industry might inadvertently have enhanced the level of fear that surrounds nuclear power, by stressing safety issues so much. However, he comes perilously close to claiming that the main culprit was what he labels the 'anti-nuclear brigade'. For them, 'the misunderstanding of radiation is an important key to discrediting nuclear power', and in this way they raise unwarranted fears. 'Yet they are the very people who are inducing such effects by continuing to feed the public scares about radiation! So the psychological impacts become essentially self-fulfilling; they stoke up an (illusory) fear and then complain about the consequences of this.'

In a subsequent article, he suggested that 'the high and apparently rising capital investment costs of nuclear plants in the western world'

DOI: 10.1057/9781137274335

might in part be due to the success of the anti-nuclear movement. He asks, 'Could it be that public acceptance issue is at least partly to blame for the high costs of nuclear; indeed, perhaps not only the capital investment costs but also the operating costs of plants?' (Kidd, 2012b).

Given that, as we have seen, the nuclear industry is clearly suffering economic problems, with Fukushima adding more, it is perhaps not surprising that some parts of the nuclear lobby have become a little shrill. But it seems odd to try to blame the opposition for their problems. It may be true that the anti-nuclear movement, which contains a wide range of groups of various types, can at times be less than rigorous in its use of campaigning arguments, but the issues it seeks to raise are, arguably, real ones, reflecting real concerns and risks, and technological choices. The relative significance of some of the risks can be debated, as can the benefits or otherwise of the various technologies, but to suggest that the nuclear lobby has a unique grasp of the truth seems, to put it mildly, unconvincing and possibly inappropriate.

DOI: 10.1057/9781137274335

8

Reactions to Fukushima: Contestation and Trust

Abstract: *In addition to the practical problems of clearing up the mess, the legacy of Fukushima includes a collapse of trust in the authorities in Japan, and in nuclear power around the world, with opposition at high levels in most countries, but governments in many still pressing ahead with nuclear expansion. Given this conflict, the media and agencies which relay views on issues such as heath and safety play crucial roles. After reviewing the scale of the opposition, this chapter looks at some contested health impact issues. For those directly affected, the debate is not just of academic interest. This chapter also looks at the broader debate about future energy options – key factors in which are the widespread opposition to nuclear and the increased attractiveness of the alternatives.*

Keywords: energy policies; media coverage; public reactions; radiation and health

Elliott, David. *Fukushima: Impacts and Implications.* Basingstoke: Palgrave Macmillan, 2013. DOI: 10.1057/9781137274335.

DOI: 10.1057/9781137274335

> To insist as a matter of science that nuclear is indispensable, is undermining both of scientific independence and healthy democratic debate.
>
> Professor Andrew Stirling, University of Sussex (Stirling, 2011)

8.1   The spread of opposition

Nuclear power is clearly not popular, but those opposed to it are faced with a battle over technological choice, with views on the risks, safety and heath impacts of nuclear power, as well as on economics, security and many other issues, being contested. The balance in the debate has shifted. German Chancellor Angela Merkel commented that Fukushima 'has forever changed the way we define risk', and certainly there are many who now do not trust the current approach to risk assessment (Dorfman, 2012; Blowers, 2012).

At present, only 30 countries, out of the 192 members of the United Nations, operate nuclear plants – about 16%. The industry clearly wants that figure to expand. But Fukushima will make that hard. Trust in nuclear power, and in those who support it, has been seriously undermined. A new anti-nuclear movement has emerged and is becoming increasingly active and confident locally and globally. Moreover, support for its views seems to be spreading.

A 24-country public opinion study carried out by Ipsos in May 2011 found that 62% of those asked opposed nuclear power. Overall, 26% had changed their mind after Fukushima, tipping the scales against the nuclear option. (See Table 8.1 for some key results.) Opposition in some developing countries and in much of Europe was clearly very high (Ipsos, 2011a).

Similar results, although with some key differences, emerged from a GlobeScan opinion poll, commissioned by BBC News, covering 23 countries and carried out between July and September 2011 (Black, 2011; BBC World Service, 2011). (See Table 8.2 for some key results.) I noted some of the more detailed individual country results from this global poll, and from the Ipsos poll, in earlier chapters.

Overall, 71% of respondents thought their country 'could almost entirely replace coal and nuclear energy within 20 years by becoming highly energy-efficient and focusing on generating energy from the

DOI: 10.1057/9781137274335

TABLE 8.1   *Increased opposition to nuclear after Fukushima (%)*

	Opposed before	Opposed after	Change
Germany	67	79	+12
Turkey	52	71	+19
Brazil	55	69	+14
France	53	67	+15
Russia	52	62	+10
Japan	28	58	+30
China	28	58	+30
UK	41	51	+10
US	36	48	+10
India	19	39	+20

Sources: Ipsos (2011a).

TABLE 8.2   *Support for nuclear before and after Fukushima (%)*

'Nuclear power is relatively safe and an important source of electricity, and we should build more nuclear power plants.'

	Those agreeing:	
	2005	2011
France	25	15
Germany	22	7
India	33	23
Indonesia	33	12
Japan	21	6
Mexico	32	18
Russia	22	9
UK	33	37
US	40	39

Sources: BBC World Service (2011), Black (2011).

Sun and wind', while only 22% agreed that 'nuclear power is relatively safe and an important source of electricity, and we should build more nuclear power plants'. Nevertheless, there was a degree of resignation to the status quo: globally, 39% wanted to continue using existing plants without building new ones, although 30% wanted them all shut. In

DOI: 10.1057/9781137274335

countries without operating reactors, support for nuclear was greatest in Nigeria (41%), Ghana (33%) and Egypt (31%). It also remained high in some countries with reactors. Support was high in China and Pakistan, at around 40%.

The headline figures, though, were that opposition to nuclear power in Germany was up from 73% in 2005 to 90%. It also rose in France (66% to 83%) and Russia (61% to 83%), while in Japan the rise was from 76% to 84%. However, uniquely, in the UK the poll found that support for new reactors had risen from 33% to 37%.

That result was not backed up by other polls looking at the initial impact. Indeed, as noted above, in its May 2011 poll Ipsos found that opposition in the UK had risen from 41% to 51% after Fukushima (Ipsos, 2011a), and in its June 2011 poll it found that support for a replacement programme had fallen from 47% (in November 2010) to 36% (Ipsos, 2011b). However, Ipsos also said that subsequently, by December 2011, UK support had reverted to pre-Fukushima levels (indeed slightly higher, at 50%), while opposition had fallen. Ipsos concluded that anti-nuclear feeling following the March 2011 accident 'now looks like no more than a temporary blip, as year-on-year improvement in support has resumed' (WNN, 2012d). A very different view of UK reactions emerged from a *Guardian*/ICM poll published in March 2012, which showed that support for building a nuclear power station near people's homes had fallen markedly since Fukushima, from 24% in 2010 to 14%, with those opposed rising from 60% to 72% (*Guardian*, 2012).

Poll data has its limits: it depends on the questions asked and their framing. But if the same questions are repeated, they can reveal trends over time that may still be significant – indicating, for example, how events such as Fukushima affect views. Clearly, in addition to variations in national views over time, there are also major international differences in the poll results, reflecting the complexity of the issues and varying local concerns and contexts. In seeking to find out what might have been influencing these results, I looked in the previous chapter at some political and economic factors, but came to no simple conclusions. Although there were interactions, they were complex. Could other influences also be at work, shaping reactions?

In this context, it is perhaps instructive to look at the US, where an opinion survey in September 2011, carried out by Bisconti Research in conjunction with GfK Roper on behalf of the Nuclear Energy Institute, found that 62% of respondents favoured the use of nuclear as

DOI: 10.1057/9781137274335

one way to generate electricity in the US. It noted that this represented a small decrease in those supporting nuclear since a similar survey in February 2011, just before Fukushima, when 71% were in favour. It also noted that 26% of those asked in February had said they opposed nuclear energy, while the new figure was 35%. In effect, it suggested, the accident had resulted in 9% of people changing their minds (WNN, 2011e).

Ann Bisconti, president of Bisconti Research, commented:

> While there is some evidence of the impact of the Fukushima events, support for nuclear energy continues at much higher levels than in earlier decades. Turmoil in oil-rich areas of the world and hikes in oil prices historically have focused public opinion even more on nuclear energy, and may have helped to preclude a serious impact of events in Japan on public attitudes.

She added, 'The information heard about nuclear power plant safety here in the US also likely helped to keep support at high levels.' What we are hearing here is that wider concerns affect views and that, in an uncertain climate, pro-nuclear information works.

In this context it may be worth noting an infamous leaked email from a staff member of the UK Department for Business, Innovation and Skills, following Fukushima. It suggested that Fukushima 'has the potential to set the nuclear industry back globally. We need to ensure the anti-nuclear chaps and chapesses do not gain ground on this. We need to occupy the territory and hold it. We really need to show the safety of nuclear' (Edwards, 2011).

8.2   The role of the media

While governments and the nuclear industry, and its supporters, have access to their own means of propagating positive views on nuclear, views are also likely to be influenced by the media, who will obviously be a target for PR efforts. It has been claimed that 'the media are saturated with a skilled, intensive, and effective advocacy campaign by the nuclear industry and its powerful allies' (Lovins, 2012).

How do assertions like this stand up in the light of Fukushima? At first glance, not very well. During and immediately after the Fukushima

DOI: 10.1057/9781137274335

disaster, there was no shortage of negative media messages on nuclear. For example, a review of UK media output following Fukushima by the Science Media Centre noted that much of the press and TV news channel coverage was alarmist, although there were some exceptions (Fox, 2011).

However, as I noted in Chapter 5, some subsequent coverage by the BBC adopted a different approach. The BBC prides itself on the balance of its output, but some people saw two of the TV documentaries it produced after Fukushima as having failed in that regard. For example, the message from the *Bang Goes the Theory* programme was that not many had died at Chernobyl and none would die from radiation at Fukushima.

There was no coverage of the range of scientific opinion on, for example, the potential negative long-term impacts of low-level radiation contamination and the significance of absorbed 'internal emitters' (CERRIE, 2004). The impacts of radiation are contested, and one might think this debate worthy of proper coverage. It certainly does get covered on the web, even if the quality of the contributions varies.

It is understandable that the media wants to simplify issues, but it is more usual for it to delight in controversy. If nothing else, that might lead to a more entertaining exploration of conflicting views.

In the written exchanges on the formal complaint by Nuclear Consult about the alleged lack of balance in the *Bang Goes the Theory* programme, the BBC said, in an interim comment:

> the divergent views and debate relating to nuclear incidents at Chernobyl and Fukushima relate primarily to long term and indirect health and environmental impacts. In making the decision to present none of these divergent views within this programme, yet consider them all in drawing our own conclusions, we sought to avoid making unfair representation of any one view. Rather we presented only statistics which have been officially reported with firmly substantiated evidence. (NGC, 2011)

It is hard to know what impact this agenda-limiting approach might have had on public opinion. These programmes were aired at peak viewing times, so they were probably quite influential, especially given the prestige of the BBC as an allegedly impartial observer. It may be that some of its other outputs were more balanced; for example, BBC Scotland transmitted what some saw as a much fairer report, and the BBC website has had some balanced coverage.

DOI: 10.1057/9781137274335

Nevertheless, repetition in other media of the message that fears about Fukushima were overblown may well have consolidated its hold. For example, the popular UK science magazine *New Scientist* seemed to adopt a pro-nuclear line. In its 26 March 2011 issue, it assigned only half a page of six full pages of coverage to information on the unfolding consequences of Fukushima. The remainder were devoted to a cause referred to in a sub-title as 'rescuing nuclear power' to deliver what the cover asserted to be 'our nuclear future'.

In a critique Professor Andy Stirling from Sussex University commented:

> Current measured, evidence-based judgements that new nuclear build does not offer a favourable element in low carbon strategies were referred to as 'wild proclamations of the end of the nuclear era'. The false assertion was made twice without substantiation, that nuclear power is an 'essential option'. This looks less like reasoned, balanced coverage of a complex, uncertain science policy issue, and more like biased propaganda: self-fulfilling prophecies in a particular sectoral interest.

He said, 'In fact, it is documented incontestably in multiple international studies of the most authoritative standing, that – with the requisite political will and investment – it is physically possible, technically feasible and potentially economically viable for the world to achieve a resource-diverse, zero carbon future over the necessary period of a few decades, without new nuclear build.' He concluded that 'support for nuclear power is entirely legitimate, but it is just one particular political position. To insist as a matter of science that nuclear is indispensable, is undermining both of scientific independence and healthy democratic debate' (Stirling, 2011).

Some might argue that the BBC and *New Scientist* were trying to balance what they thought were alarmist portrayals of Fukushima from the anti-nuclear movement and some of the more lurid popular media coverage. Certainly that was the line adopted by the *Bang Goes the Theory* programme, which concluded its presentation of estimates of death at Chernobyl by saying, 'Figures like these certainly suggest that radiation from accidents like Chernobyl is not as worrying as a lot of the media coverage would have us believe.' Adopting a similarly calming tone, *New Scientist*, in its 16 November 2011 issue, quoted the view of Professor Gerry Thomas at Imperial College that we did not need to worry too much about Fukushima since 'not an awful lot got out of the plant – it was not Chernobyl'.

While it may be true that some of the mass media can be alarmist, sensationalist and shallow in its approach, there is surely a role for more

DOI: 10.1057/9781137274335

thorough and open analysis of the issues. In such a polarised situation that may, of course, be hard. Like the popular media, the anti-nuclear movement may sometimes be guilty of alarmism, although perhaps this is not surprising given that its views, and those of experts who are critical of nuclear, rarely receive much serious media coverage. To get attention, they have to render their views in dramatic terms or organise public demonstrations and actions under simple slogans.

Since there is not much coverage of their more developed views, at least in the main public media, we do not know what effect they might have on public opinion. Nevertheless, the nuclear lobby evidently sees anti-nuclear views as a threat and expends effort countering them. It may be misguided in its assessment of the impact of anti-nuclear views, although clearly anti-nuclear views do have currency.

In this context, it may be a wry (and amusing) indication of the apparent strength of fear of the influence of anti-nuclear views that a complaint was made by the president of the Royal Society of Chemistry in January 2012 about negative images of nuclear power in James Bond movies such as *Dr. No.*

More seriously, at one point the Japanese authorities, evidently concerned about attempts to 'stir up mischief' and 'create public anxiety' via 'false rumours' on the internet and in other media over contamination risks and safety issues, asked telephone companies, internet service providers, cable television stations and others 'to take the necessary measures' while 'giving consideration to freedom of expression' (MIC, 2011). This opens up some interesting questions about the role (and responsibilities) of social media and journalism generally, especially during times of crisis.

In terms of the wider, ongoing public debate about nuclear power, since the nuclear issue is a contested one it is understandable that all those involved will seek to put their views forth strongly, in any way possible. The media (press and TV) clearly play an important role in reporting on this debate, as well as on key events. As we have seen, there has been some limited discussion of the role of the media, broadly defined, in the recent nuclear debate. There were some discussions of this issue after Chernobyl (Gamson and Modigliani, 1989). We can now expect more.

Thankfully, the media coverage around the time of the first anniversary of the Fukushima disaster was generally more balanced. Perhaps wisely, most TV programmes (including an informative documentary produced for BBC 2) were limited to reprising or providing new accounts of the events, rather than offering analysis. However, the UK

DOI: 10.1057/9781137274335

Guardian newspaper, in an editorial at the end of February 2012, came off the fence: the UK, it said, 'must not assume the same thing couldn't happen here. That's what they said in Japan after Chernobyl.' The usually very conservative *Economist* magazine also seemed to adopt a critical line, with a feature entitled 'The Dream That Failed' (Morton, 2012).

8.3   Whom to trust?

In the debate over nuclear power, we tend to rely on those sources we trust. However, it is sometimes hard to know whom to trust. Certainly the media can be unreliable, as can some sources on the internet. The level of trust in governments on the nuclear issue has fallen to an all-time low in many (but not all) countries, fuelled by what many people see as a failure by the Japanese authorities to present timely and accurate reports of what was happening and the risks. In the Ipsos poll, 28% of the Japanese asked did not think Japanese officials and government leaders had communicated the nature and impact of the accident to the Japanese people and others honestly, and the percentage of respondents holding this critical view was very much higher in some neighbouring countries, reaching 89% in Indonesia and 90% in India (Ipsos, 2011a). Critical views also emerged in relation to other governments. For example, in France an Institut de Radioprotection et de Sûreté Nucléaire survey in September–October 2011 found that only 24% of their sample said they would 'trust the authorities' on the risks issue (IRSN, 2012).

Nevertheless, some international agencies may still be trusted. For example, for clarity on nuclear health and safety issues, many will look to WHO, the UN's World Heath Organization. How independent is it? In 1959, WHO signed an agreement with the IAEA which says, 'Whenever either organisation proposes to initiate a programme or activity on a subject in which the other organisation has or may have a substantial interest, the first party shall consult the other with a view to adjusting the matter by mutual agreement.'

The IAEA combines nuclear assessment and regulation with strong nuclear promotion. This, arguably, is problematic. There are major disputes about the impacts of radiation. The standard, widely accepted view is that the biological impacts of radiation increase linearly with intensity of exposure. However, there are rival views. Some of these claim that low levels are less harmful than this linear model suggests (Wade, 2009);

DOI: 10.1057/9781137274335

unsurprisingly, the industry is attracted to such views. A diametrically opposed view, backed by a number of reputable scientists, is that the current linear model may underestimate actual risks (Brenner *et al.*, 2003).

Issues such as this, which have implications for longer-term impacts, were avoided in a recent report on Chernobyl by the UN Scientific Committee on the Effects of Atomic Radiation, published in February 2011. It said that the known death toll from Chernobyl was 28 fatalities among emergency workers, plus 15 fatal cases of child thyroid cancer by 2005. It added, 'To date, there has been no persuasive evidence of any other health effect in the general population that can be attributed to radiation exposure.' Crucially, it did not speculate about future deaths 'because of unacceptable uncertainties in the predictions' (UNSCEAR, 2011).

How reliable is this assessment? It seems odd to avoid discussing longer-term impacts. Many independent scientists consider UNSCEAR's conclusions to be unreliable, and some have called for further monitoring work to clarify longer-term outcomes (ARCH, 2012). Certainly there are studies suggesting that the ultimate death toll from Chernobyl could be very much higher than that predicted officially. Radiation exposure is now thought to have an impact on the cardiovascular system and can compromise the immune system. Since it can take time for any ill-effects to show, there may be health impacts later which are not attributed to earlier nuclear contamination.

Many reports of high levels of illness persist in affected areas of Russia, Ukraine and Belarus. Drawing on some of this data, a study by researchers from the Centre for Russian Environmental Policy and the Institute of Radiation Safety in Belarus, published by the New York Academy of Sciences, claimed that almost one million people around the world may have died from exposure to radiation from Chernobyl (Yablokov *et al.*, 2009).

This estimate may be considerably exaggerated. Some critics have claimed that there were methodological and other weaknesses in the study (Balonov, 2011; Charles, 2010; Jargin, 2010). In this situation it is tempting to take the middle ground. Certainly what seems to be a balanced, independent study of Chernobyl, based on survey data from across the whole region and using extrapolations from standard dispersion models, estimates the ultimate regional death toll at 30,000–60,000 (Fairlie and Sumners, 2006). As for Fukushima, we do not yet know. The short-term impact seems thankfully to have been low, but there have been some provocative estimates suggesting there might ultimately be thousands of future deaths (Busby, 2011).

DOI: 10.1057/9781137274335

While many will discount estimates like these as extreme, the debate over impacts and risks will no doubt continue. To some extent this is an irresolvable issue. Perceptions of risk and its significance will often be shaped by philosophical or perhaps political views. If, for example, a large group of people are exposed to a low level of radiation, each individual may see his or her risk of contacting cancer as small: the chance is low. But, viewed collectively, some people will be unlucky, and it is collectively that we seek to reduce risk.

For some people any increase in exposure should be avoided. Others may say that life is risky and that we cannot remove all risks. Yet the risks with nuclear are inevitably much higher than with, say, wind or solar power. While debate continues in the abstract, in practical terms the focus may be on how public and occupational safety levels are set. The nuclear industry is likely to want the levels to be set higher, to reduce operating costs.

This issue came to the fore in Japan after Fukushima. As the *Financial Times* reported in May 2011, a radiation safety researcher, Professor Toshiso Kosako, resigned as a scientific adviser to Japan's Prime Minister after the government raised the limit for exposure in schools from 1 mSv a year to 20 mSv a year, a level usually used for nuclear industry workers. The *Financial Times* quoted her as saying, 'It's unacceptable to apply this figure to infants, toddlers and primary school pupils' (Dickie and Cookson, 2011). However the paper also quoted Professor Wade Allison of Oxford University, who said the 20 mSv a year threshold for evacuation should be raised to 100 mSv a month (i.e. 1,200 mSv a year). He claimed that the principal health threat posed by the Fukushima Daiichi crisis was 'fear, uncertainty and enforced evacuation'.

This is clearly a controversial view, and one evidently not shared by many radiation scientists (Fairlie, 2010). Moreover, the issue goes beyond just exposures after accidents: there is also the issue of the impact of permitted routine emissions of nuclides. There have been reports claiming elevated levels of leukaemia in children living near nuclear plants. For example, the 2008 German government KiKK study found large increases in leukaemia (120%) and embryonal cancer (60%) among children living near all German nuclear reactors (Fairlie, 2009). Similar results have emerged from a recent study in France, which found a statistically significant doubling of the incidence of leukaemia near to nuclear plants (Sermage-Faure *et al.*, 2012).

Official agencies continue to insist that the dose estimates from nuclear emissions are too low to result in the observed levels of leukaemias.

DOI: 10.1057/9781137274335

Certainly it is hard to prove a causal link between gaseous discharges from plants and ill-health: there could be other causes. The UK scientific advisory body the Committee on Medical Aspects of Radiation in the Environment has, for example, pointed to unidentified viral infections, rather than radiation exposure, as a possible explanation for the results from the KiKK study (COMARE, 2011).

Clearly scientific disagreements remain over the health effects of radiation, with unresolved disputes about the dose effect model that should be used (Safegrounds, 2011). Tragically, the European Commission and WHO appear to have shied away from funding a study that might have provided a clearer understanding of what the risks are, based on looking at lifetime impacts from exposures to radiation from Chernobyl (Butler, 2011; see Section A.3 of the appendix for more discussion).

Meanwhile, as many Japanese nuclear workers and residents worry about the longer-term impacts of Fukushima, Japan cannot afford to wait for a resolution to the debate over radiation and health issues, or for the results of further research. In addition to trying to make the wrecked reactors safe, it has to face the huge, urgent, practical problems of clearing up and dealing with the aftermath of the Fukushima disaster and the social and economic dislocation that resulted from it (Takano and Takano, 2012).

The collapse of trust in the authorities may make this process much harder. Many of the thousands of people who were displaced are trying to re-establish their lives, but, as we have seen, they often do not know whom to believe. They also face bureaucratic obstacles.

There was talk of victims receiving $7,000 each in compensation, but TEPCO has evidently made it hard for them to apply: applicants have to fill out a 60-page compensation claim form, which is accompanied by a 160-page explanation booklet. Claimants have to attach evidence about properties and assets, documents which many lost in the quake and tsunami that destroyed their homes or which remain at home, in the exclusion zone. Such is the loss of trust that many victims suspected TEPCO was making it deliberately hard to claim to deter all but the most dedicated applicants (McNeill, 2011b).

The clean-up operation itself will not be easy, especially given the breakdown in trust. Sites have to be found for burial of the radioactive material cleared from contaminated areas. One aim is to landfill it, using plastic sheeting for lining. However, there have already been local conflicts over this, and over burning contaminated wastes from the area in incinerators. The Japanese government seems to see local municipalities as being responsible for dealing with some of this waste, which has raised tensions further (Chillymanjaro, 2011).

8.4 Fukushima's legacy

After Fukushima, Japan clearly faces major problems. Tragically, it already had experience of a major nuclear crisis, following the bombings of Hiroshima and Nagasaki, but it recovered. Given that the scale of the impact of Fukushima is still contested, it may be instructive, if harrowing, to compare the aftermath of Fukushima with that of the US nuclear bombings in 1945. The differences are vast. The attacks led to massive loss of life, both immediately and afterwards, due to the blast and heat flash, direct radiation exposure and subsequent fallout.

However, apart from the initial gamma and neutron flash, the radioactivity was largely limited to the residual mass of the by-products from the fission of a few kilograms of uranium (Hiroshima) or plutonium (Nagasaki) and the activated bomb components.

These were air-bursts at 500–600 m altitude, designed to create the maximum blast damage. Therefore, although there was radioactive rain, the resultant radioactivity in soil and debris at ground level was relatively low. Most of the fallout also fell outside the cities because of prevailing wind patterns.

Although there is no diminishing the horrendous scale of the immediate and continuing human impacts of these weapons, both cities were deemed (admittedly, perhaps prematurely by modern standards) to be re-inhabitable soon after and have since become thriving metropolises (Hiroshima City and Nagasaki City, 1981).

By contrast, at Fukushima only a handful of people were killed by the explosions and as yet none as a result of the released radioactivity, although, of course, there are long latency periods before most cancers appear. However, compared with Hiroshima and Nagasaki, it could take a long time to deal with the contamination legacy of Fukushima.

There were 4,277 tonnes of nuclear fuel in the Fukushima reactors and spent fuel stores. A report issued by Prime Minister Naoto Kan's Cabinet, just before he stood down, estimated that the amount of radioactive caesium-137 released at Fukushima by that time was equal to 168 Hiroshima bombs – 15,000 TBq, compared with 89 TBq released at Hiroshima (*Telegraph*, 2011b).

DOI: 10.1057/9781137274335

As we have seen, comparisons have also been made with Chernobyl, which is perhaps a more relevant exercise. We do not yet know exactly how much of the total radioactive materials inventory was released at Fukushima. It was presumably only a relatively small fraction of what was there, especially compared with Chernobyl, where the core exploded and vented to the air. However, the Chernobyl plant had only 180 tonnes of fuel – much less than the 4,277 tonnes in the reactors and waste stores at Fukushima. As a result, as noted earlier, it has been estimated that, with several reactors and waste stores involved, the total release from Fukushima to air and sea was similar to that at Chernobyl in net terms. Although only some of that was long-lived isotopes, and although some of the airborne plume went out to sea, the clean-up problems on land will still be immense.

At Chernobyl, while efforts were made to secure the reactor itself, it was decided to permanently evacuate a zone with a 30 km radius from the plant, rather than attempt what would be a very expensive exercise in remediation of soil and buildings. More than 115,000 people were relocated from this area, and, later, about 220,000 more were relocated from other contaminated areas in Belarus, Ukraine and the Russian Federation (UNSCEAR, 2008). By contrast, because land is scarce in Japan, it has been decided to attempt a full clean-up of both the plant and the affected areas around it. This will be difficult and expensive.

It will be decades before the reactors, associated buildings and spent fuel tanks at Fukushima are fully secured. The massive land decontamination process will also probably take decades. It will include remediation of farm-land and forest areas. Japan's Science Ministry said that 8% of the country's land area was contaminated with caesium-137 at various levels. An estimate by Asahi news service suggested that around 13,000 km^2 of land across eight prefectures would have to be decontaminated (Ishizuka and Mori, 2011). In the worst areas, top-soil is being removed and leaf litter cleared from woodland, around 29 million m^3 in all.

However, there are limits to how much can be done without causing ecosystem problems. A decontamination expert from the Japanese Atomic Energy Agency commented, 'You remove leaf litter from the forest floor and radiation levels fall. You take away the deeper layers and they fall more. But you take it all away and the ecosystem is destroyed. Water retention goes down and flooding can occur' (Bird, 2012).

That should remind us that it is also possible that some of the contamination will seep into the soil and end up in the water table;

DOI: 10.1057/9781137274335

remediating the land may be only part of the problem. In addition to land and groundwater contamination, over 110,000 tonnes of contaminated water was pumped or leaked into the sea during the attempts to cool the Fukushima Daiichi reactors, and this may also have longer-term implications. France's nuclear monitor, IRSN, said that the amount of caesium-137 that leaked into the Pacific from Fukushima constituted the greatest single nuclear contamination of the sea ever.

Although the contamination will be hugely diluted by ocean currents, 'significant pollution of seawater on the coast near the damaged plant could persist' as a result of continuing runoff of contaminated rainwater from the land. So 'maintaining monitoring of marine species taken in Fukushima's coastal waters is justified'. IRSN listed deep-water fish, fish at the top of the marine food chain, molluscs and other filtrating organisms as 'the species that are the most sensitive' to caesium-137, which has a 30-year half-life (IRSN, 2011).

As noted earlier, there were also concerns about the contamination of other food. In one notorious example, meat from cattle that had been contaminated by their feed was passed into the food chain (Japan Focus, 2011). It seems that the cows were kept indoors to protect them from fallout, and monitored externally, but their ingestion of contaminated hay was initially overlooked (Yomiuri, 2011). As a result, hay contaminated with caesium at levels up to 690,000 Bq/kg, compared with a government safety standard of 300 Bq/kg, was evidently fed to cattle. It appears that 4,108 kg of beef suspected of being contaminated was inadvertently put on sale at 174 shops across Japan before the authorities realised there was a problem. In July 2011, the Health Ministry was reported to have found caesium-137 contamination at 2,300 Bq/kg (Takada, 2011). Such issues are presumably one reason tourism, an important industry in Japan, has suffered, falling by 60% immediately after the accident and remaining low thereafter.

Export of foodstuffs around the world, another key industry for Japan, has also been hit. A global Ipsos poll in May 2011 found that 45% of their world sample had avoided at least one Japanese product because of Fukushima, with consumers in China and South Korea being particularly sensitive (Ipsos, 2011a). There were also worrying incidents of negative reactions to Japanese travellers. In May, China detained two Japanese visitors who had arrived by air for radiation checks. They apparently had contamination on their clothes and shoes.

As noted earlier, there were initial concerns about fallout from Fukushima spreading to other countries in the region. That had certainly

DOI: 10.1057/9781137274335

happened at Chernobyl when the radioactive plume went westwards, covering most of Europe, which led to major and continuing contamination issues. For example, the sale of lamb from some sheep farms in Wales is still subject to control orders to this day.

Because of the mostly fortunate pattern of wind dispersal, we might expect fewer problems from Fukushima outside Japan, but there could still be some, and the marine impacts from the release of radioactive water could also have a wider reach. For example, China expressed concern about possible contamination of its waters (CNTV, 2011).

While contamination issues are important, clearly the reactions to major accidents like this are likely to go beyond health and safety issues to wider economic and political concerns. For example, in addition to the health issues and their economic impacts, another potentially important legacy of Fukushima is the strengthening of concerns about the viability of nuclear power around the world. That was one result of Chernobyl, and it also seems to have been a major response to Fukushima. While heath and safety issues obviously worry those opposed to nuclear power, many are also concerned by what they see as the economic madness of investing in costly technology which relies on a fuel that is finite and whose use creates waste to guard for millennia. For others the concern is the weapons proliferation risk or the threat of terrorist attacks on nuclear facilities.

There are many other technical, economic, security and strategic issues (Elliott, 2010), and in the aftermath of Fukushima some, such as the economics questions, are resurfacing. Indeed, one reason that some people in the UK do not worry unduly about nuclear is that they do not believe that any new plants will actually get built, given the escalating costs and the parlous state of EDF's finances. This may be to ignore the UK government's determination to support nuclear expansion by all means possible, including radical and complex rearrangements of the UK electricity and carbon markets.

8.5   What happens next?

Fukushima is likely to remain a focus of concern for some while, most obviously in Japan. It will take a long time before the displaced people can re-establish their lives, and even longer before a full clean-up operation can be completed. Health impacts may also begin to emerge. There

DOI: 10.1057/9781137274335

is also the risk of another major earthquake, and even a tsunami, in Japan, which the crippled plant at Fukushima would be in a poor state to withstand. But leaving all that aside, debate over nuclear power in Japan and elsewhere is likely to remain intense, given that, as we have seen, many issues are still contested.

While some will continue to see nuclear power and its advocates and supporters as untrustworthy, others will adhere to the hope that the technical problems with nuclear may be resolvable. Certainly there are those who look to a continuing future for nuclear power, based perhaps of the use of thorium in molten-salt breeder reactors – or, even further ahead, to fusion (see Section A.2).

There are even some who see Fukushima as proving that nuclear is safe, or at least safe enough; after all, they say, the plants managed to survive the worst conceivable accident without releasing massive amounts of radioactive material (Monbiot, 2011). Most people, however, are likely to see Fukushima as a wake-up call, alerting us to the need to think again about nuclear.

Although renewable energy has its detractors, many of whom are nuclear proponents, the case for rapid deployment gets stronger by the day. In a report on 'Deploying Renewables' the International Energy Agency said that 'a portfolio of renewable energy technologies is becoming cost-competitive in an increasingly broad range of circumstances, in some cases providing investment opportunities without the need for specific economic support'.

It now seems realistic to think of obtaining nearly 100% of electricity, and perhaps also of energy, from renewable sources by around the middle of the century in much of the world, given the right levels of support and backing for energy efficiency (WWF, 2011b; Jacobson and Delucchi, 2009, 2011).

Obviously renewables have their problems, but these are mostly operational issues related to managing their variability, very different from the safety, security, waste and proliferation problems associated with nuclear, and relatively easy to deal with.

The grid system, after all, already deals with large daily swings in demand and occasional closures of power plants (including nuclear plants), and there are many other technical options for providing the necessary balancing, at a small extra cost (Boyle, 2007).

Some say we must back both nuclear and renewables, but that risks diluting financial and technical resources, and doing neither well.

DOI: 10.1057/9781137274335

Moreover, it has been argued that nuclear and renewables are technically and strategically incompatible (Froggatt, 2010; IWES, 2010). For example, there are operational conflicts: nuclear cannot easily back up variable renewables, and grid systems would be hard to manage with large amounts of basically inflexible centralised nuclear and variable decentralised renewables both trying to feed in. This conflict was evidently recognised by EDF, which claimed that in the UK context, 'as the intermittent renewable capacity approaches the Government's 32% proposed target, if wind is not to be constrained (in order to meet the renewable target), it would be necessary to attempt to constrain nuclear' (EDF, 2008). The point is that, if there are large amounts of both on the grid (e.g. 20 GW of each), one or other has to give way each time energy demand falls below their combined output, and especially when it goes below base-load requirements (around 20 GW).

While support for renewables is strong and growing, it has been said that nuclear needs climate change more than climate change needs nuclear. After Fukushima it might be the case that we do not want, and actually do not need, nuclear at all. Globally, the number of operating reactors fell from 441 at the start of 2011 to 435 in early 2012, a decrease of around 10 GW or 3%, mostly as a response to Fukushima, with total output in 2011 falling by 4.3% (WNN, 2012e). Some will welcome that trend.

As we have seen, Fukushima led Japan to reverse its nuclear policy, as did Malaysia and the Philippines. Taiwan also now seems committed to moving away from nuclear, and Bahrain and Kuwait have decided not to go down the nuclear route. Even more dramatically, Germany is phasing out nuclear, as are Belgium and Switzerland, while Italy has abandoned its nuclear programme. France may well start a phase-out. This certainly looks like a major shift in some key market areas.

However, the industry may well recover in other areas. In its 'Energy Outlook 2030', published in January 2012, BP says that although 'growth is concentrated in renewable power', nuclear output will be 'restored to pre-Fukushima levels by 2020, but thereafter shows only modest growth'. But in non-OECD countries growth will be 'more evenly split between renewables, nuclear and hydro, as rapidly growing economies call on all available sources of energy supply' (BP, 2012).

Overall, with world energy demand projected to grow by 39% over the next 20 years, nuclear continues to expand in the East (notably in China and India) and also in Russia. BP says nuclear output will reach 5.3% of total global primary energy output in 2020, and 6.1% in 2030 (BP, 2012).

DOI: 10.1057/9781137274335

In a report marking the first anniversary of the Fukushima disaster, the World Energy Council (WEC) was even more bullish. It commented, 'Very little has changed ... in respect of the future utilisation of nuclear in the energy mix.' After surveying its members in 94 countries, according to senior project manager Ayed Al-Qatani, the WEC found that '[t]he Fukushima accident has not led to any significant retraction in nuclear energy programs in countries outside Germany, Switzerland, Italy and Japan'. Progress in some countries had been delayed, but there was 'no indication that their pursuit of nuclear power has declined in response to Fukushima' (WEC, 2012; WNN, 2012e).

As we have seen, that may not be entirely accurate, and certainly does not reflect the rising level of opposition around the world. However, the World Nuclear Association went on the offensive and asserted that 'people can draw confidence from the absence of any health harm even from this extreme, highly unusual event'. Its director general, John Ritch, claimed that 'countries like Germany will soon demonstrate the economic and environmental irresponsibility of allowing politicians to set important national policies in the middle of a panic attack' (WNN, 2012f).

So far such predictions do not seem to have proved correct. Although utilities such as E.ON were hardly likely to support the German nuclear phase-out, they seem to have adopted a new approach, at least in terms of new plants, even in relation to projects in the UK, where there was a government clearly keen to support nuclear. Following E.ON's withdrawal from the Horizon nuclear project in the UK in March 2012, its chief executive officer was quoted as saying that, while investment in nuclear could still be profitable, 'our priorities have changed. ... We have come to the conclusion that investments in renewable energies, decentralised generation and energy efficiency are more attractive – both for us and for our British customers' (Teyssen, 2012).

DOI: 10.1057/9781137274335

9

Conclusions: The Lessons of Fukushima

Abstract: *There are practical lessons from Fukushima, but some more general issues also emerge concerning what happens following major accidents like this, the debate over nuclear safety and the wider debate about energy options. It is not possible to predict when another such accident will occur or to predict what the reactions will be, but this chapter summarises what has emerged from the foregoing review and analysis of the reactions to Fukushima. It concludes that safety issues will remain paramount, recriminations over institutional failings will continue and opposition may grow, but that economic and strategic issues, and the advent of more attractive energy options, may dominate the future.*

Keywords: energy policy prospects; institutional blame; lessons of Fukushima

Elliott, David. *Fukushima: Impacts and Implications.* Basingstoke: Palgrave Macmillan, 2013. DOI: 10.1057/9781137274335.

> The Fukushima nuclear accident exposes the deep and systemic failure of the very institutions that are supposed to control nuclear power and protect people from its accidents.
>
> Greenpeace International, 'The Lessons of Fukushima' (Greenpeace, 2012)

9.1   The future of nuclear power and reactions to it

Although some commentators say nuclear expansion may recover after Fukushima, this is not certain. There are many reasons why some people have been concerned about nuclear power, but Fukushima has raised safety concerns to a new level, with opposition to nuclear becoming widespread across the world.

Professor José Goldemberg has estimated that the probability of a major accident happening at any of the currently operating nuclear reactors over the next 20–25 years is 1 in 5,000. This means that major nuclear accidents can be expected to occur every 20 years. On the basis of earlier estimates, he says, 'we were expecting one accident over a 100-year period' (Goldemberg, 2011).

For many people, Fukushima made clear, once again, just how risky nuclear power can be. An overwhelming disaster was fortunately avoided, but it may have been a close-run thing, and the scale of what could happen, if not this time then next, became clear to many. It is hardly surprising that, after peering over the edge of the abyss, Prime Minster Kan, on leaving office, said, 'I would like to tell the world that we should aim for a society that can function without nuclear energy.' The poll data from around the world suggests that many people agreed with him: a clear majority opposed nuclear power, although responses differed from country to country. For most people, Fukushima focused attention on the problems of nuclear power in a way that economic and other concerns had not.

It might be argued that these negative responses were unduly pessimistic. Indeed, as I have noted, some commentators saw Fukushima as demonstrating how safe nuclear power was. Despite a massive tsunami, there had been no major radioactive releases. However, as the polls and public reactions indicated, for most people Fukushima led to a major loss of trust in the technology and in those who managed it. An apocalypse had been avoided but, for many, what had happened was bad enough.

Radiation exposure is obviously an issue of grave concern to many people, and even if the risks can be overemphasised, they are sufficient to change behaviour and attitudes. While the scale of the health impacts that may result from Fukushima, and other such accidents and leaks, may have been overstated in some reports, a key issue is that, despite assurances from the authorities and from the media, many people still feel very concerned.

Some scientists have concluded that the risks from nuclear accidents of this sort, and from nuclear power generally, are low. For example, the *Financial Times* quoted the view of Dr Richard Wakeford, a radiation epidemiologist at Manchester University, who said, 'A lot of people would think it obvious that we'd be able to see health effects from an accident like Fukushima. In fact, it is not clear that we'll detect anything beyond normal fluctuations' (Dickie and Cookson, 2011).

However, as we have seen, there are differing views about the risks of radiation. It may be that few deaths will result from Fukushima. But we do not know for sure, or what might happen following any future major accidents – or, indeed, as a result of other leaks and releases. That leaves the non-expert in a quandary, especially as, whatever the risks, children and pregnant woman are the most susceptible. It is the uncertainty that for many is the major worry; while we may face many other risks, fear of radiation has an immense psychological impact, especially if exposure is involuntary.

That said, some people are complacent about, or resigned to, risks like this, seeing them, perhaps fatalistically, as part of modern life: there are too many risks to spend time worrying about them all, especially when, individually, the chance of being seriously affected by some of them, such as nuclear accidents, seems low.

It could be that, as time passes, and if few post-Fukushima health impacts emerge or are reported, opposition will fade, especially if the authorities and the media provide comforting messages about heath and safety. So, if there are no further major problems with nuclear, or if the authorities and the media manage to deflect attention from them, and perhaps also portray renewables as unreliable or costly, then, in a context where fear about climate change is growing, opposition to nuclear could dissipate. However, for the moment, with the social, political and economic fallout from Fukushima still very apparent and continuing, in many places in the world that does not seem likely. Views have changed. For many people nuclear is now not the answer to climate change, whereas the ongoing development of renewables offers hope for what many see as a better approach.

DOI: 10.1057/9781137274335

The robustness of this view is hard to determine. Fukushima clearly strengthened it, and, for those so inclined, the coverage of nuclear and energy issues on the internet from a critical perspective may ensure that it is maintained. For those who do not trust or accept the official views, the web may be the only viable source of information, even though the quality of web contributions varies. That said, some websites will be pro-nuclear; there is no shortage of well-funded pro-nuclear sites, including a substantial (and useful!) one run by the World Nuclear Association. So it is hard to decide whether the internet will make an overall difference.

What would clearly make a difference is if, tragically, there were another major nuclear accident or a terrorist attack on a nuclear facility. That would consolidate opposition. Certainly a major disaster and core release, with unfavourable wind patterns spreading significant fallout across urban areas, perhaps in nearby countries as well, would have unimaginable social, economic and political implications locally and internationally. It would probably herald the end for nuclear worldwide.

9.2 Some possible outcomes from future accidents

By their nature accidents are impossible to predict, and it is also hard to predict possible outcomes, but to try to summarise what might be learnt from Fukushima, we can at least list some of the possible outcomes from further major accidents on a similar scale to Fukushima:

1 Opposition to nuclear may grow in some but not all countries.
2 In some countries, the opposition may be sufficient to force changes in government policy, particularly if the government is already unpopular.
3 Organisational changes will usually follow; some may be cosmetic but some may be more radical, to appease the opposition.
4 Minor, or in some cases major, technical changes are likely to follow, to increase safety, or at least the appearance of safety, to appease the opposition, but also to calm the fears of existing and potential investors in nuclear projects.
5 In some cases, more effort will be devoted to developing radically new nuclear technologies for the medium or long term.
6 In some cases, the objections will be sufficient that nuclear is abandoned, or downplayed, in favour of other energy options.

DOI: 10.1057/9781137274335

7　In some cases, even if there is strong opposition, governments may persist with nuclear in the expectation that the opposition can be resisted and will fade away over time.

8　In some cases, the opposition will be weak and the issue will not need much government attention, other than issuing reassurances, with the media perhaps playing a major role.

In all but number 6 above, the cycle is likely to repeat again, when and if another accident occurs. If a roughly 20-year cycle is assumed, then the lessons from the previous incident may have been forgotten. It is conceivable, though, that concerns about climate change and energy security will reach a level at which radical action on renewables and energy efficiency is taken, to the extent that it is clear that nuclear is not needed. Conversely, a nuclear technology breakthrough offering safer and cheaper power might put nuclear more centrally on the agenda. For the moment, however, the nuclear renaissance does seem to have been stalled in much (but not all) of the world. Whether that will remain the case depends on events in the future.

9.3　The energy policy lessons of Fukushima

We can speculate about possible future events, but in terms of the actual impacts from Fukushima so far, as we have seen, while public reactions have been fairly consistent around the world, government responses have varied. In Germany, existing plans for a nuclear phase-out were reinstated and an ambitious programme of renewables expansion was launched, building on the already well-established programme. Overall, what some saw as a crisis has in fact been turned into a positive opportunity, with, so far, emissions continuing to fall. In Japan, by contrast, in the absence of a significant established renewables programme, the forced closure of all its nuclear plants has meant that there are likely to be problems, with emissions rising as fossil fuel use has increased. Hopefully, this will be a temporary phase, while a renewables programme is developed. Assuming a retreat to nuclear is avoided, the longer-term prognosis for a new approach looks positive in both Germany and Japan, with widespread backing for a shift in approach and the incumbent nuclear-oriented energy companies now on the defensive. Indeed, in Germany these companies seem to have accepted the phase-out and have switched

DOI: 10.1057/9781137274335

emphasis to renewables, while Japan may yet establish itself as a new centre for green energy innovation (Mitchell *et al.*, 2012)

While some countries in the EU and elsewhere have adopted similar approaches, in the UK the response has been very different: the nuclear push has continued, although it has faced growing problems. Nuclear power is being challenged in France, increasingly on economic grounds, leading to potential knock-on impacts for the UK programme, with investment in its proposed nuclear projects looking increasingly risky.

The economic issue has certainly become an increasing focus in critical anti-nuclear commentary (Burke *et al.*, 2012; Thomas, 2012; Toke, 2012). In September 2011, the *Financial Times* carried an article claiming that 'even before the Japanese earthquake and tsunami of March 11, prospects for nuclear construction were looking difficult in most of the developed world, mostly because of shaky economics' (Crooks, 2011). The economic and investment situation has not improved since then, with E.ON and RWE pulling out of the UK Horizon programme, and although there are reports that Chinese investment may rescue the Horizon project, concerns are also being expressed about whether GDF Suez and Centrica will continue to support other projects in the UK nuclear expansion programme. It could be that the view expressed in the redoubtable *Economist* magazine that nuclear power was 'The Dream That Failed' will turn out to be correct, at least for Europe (Morton, 2012). While it remains to be seen whether policies will change elsewhere, with the US, China, India and Russia still pressing ahead to various extents, it is clear that Fukushima, like Chernobyl before it, has led to major changes in attitudes around the world and, for Japan and several other key countries, a fundamental shift in policy.

There are, of course, many practical lessons to be learnt from Fukushima, about reactor design and location, emergency planning and so on, assuming, that is, that we continue to use nuclear power. But for many people the main lesson may be that a nuclear future is inevitably a risky one. Some of the risks are due to the nature of the technology, although new technology could perhaps reduce them. However, some of the risks are due to the nature of the institutions that design, manage and regulate the technology, and many people clearly feel that they cannot be trusted. Certainly there have been some very critical reports on the handling of the accident and its aftermath (Funabashi and Kitatazawa, 2012).

There have also been ongoing criticisms of the institutional and regulatory arrangements in Japan and in particular the relationship between

DOI: 10.1057/9781137274335

the corporations and the government, which have been heightened by Fukushima (Carpenter, 2012). Greenpeace has claimed that '[t]he Fukushima nuclear accident exposes the deep and systemic failure of the very institutions that are supposed to control nuclear power and protect people from its accidents' (Greenpeace, 2012).

Given perceptions like this and the generally negative view of the technology that many people now share, it may be that governments around the world will find it increasingly hard to ignore public opposition, especially also given the rise and appeal of the alternative, less risky and more sustainable renewable energy technologies and the economic problems that face nuclear power.

The Fukushima story of course continues, with new studies of impacts being published and new policy initiatives emerging. The text above was completed just over a year after the accident. I have provided a brief update on subsequent developments, along with some comments on how this book was written, in the Credits and Afterword section, which also includes suggestions for follow-up information sources.

DOI: 10.1057/9781137274335

Appendix: Nuclear Technology and Its Heath Impacts

A simplified guide to the basic nuclear reactor types, energy and radiation units and the impacts of radiation, with some examples from Fukushima.

▶

DOI: 10.1057/9781137274335

> Nuclear reactors are basically very complicated kettles.
>
> Anon

A.1 Energy and power units

Power ratings for power stations (and for energy-using systems) are given in watts and multiples of watts – kilowatts (1,000 watts), megawatts (1,000 kW), gigawatts (1,000 MW). The electricity produced by power stations (or consumed by users) is power × time (i.e. watt-hours, and multiples such as kilowatt-hours, kWh, the unit by which it is usually sold). Moving up the scale, a terawatt-hour (TWh) is 1,000 GWh.

Electricity is, of course, only one of the forms of energy that we use. Typically it represents around one-third of total energy use, the rest being used in the form of heating fuels and transport fuels.

A.2 Reactor types, fuels and operational issues

Most of the world's nuclear plants are based on the US pressurised water reactor (PWR) design. Others, such as the boiling water reactor (BWR) and various Russian designs, are less common. Some newer, upgraded versions of the PWR are now emerging, such as the French EPR and the US AP1000, with the scale of energy production moving up to around 1,600 MW. In addition, some smaller mini-reactor designs are emerging, in the 20–300 MW range.

In PWRs the high pressure means that the boiling point of water is raised, which makes cooling more efficient. But it also makes the reactors susceptible to rapid, catastrophic loss of cooling if the pressurised water system ruptures. BWRs, as used at Fukushima, do not face quite the same problem, since the water simply boils, unpressurised, in the reactor core, but since cooling is less efficient, they need more water throughput – a key issue at Fukushima.

Whatever the specific design of reactors, the basics of their operation are the same. A rare component of uranium ore, the isotope uranium-235 (U-235), is the only naturally occurring isotope that can sustain a chain reaction of nuclear fission which produces large amounts of heat and radiation. The heat can be used to raise steam to drive electricity-producing turbines, as in a conventional power station. The radiation

DOI: 10.1057/9781137274335

has to be contained, and the highly active waste by-products have to be dealt with – for millennia.

Plutonium-239 (Pu-239), another radioactive fissionable element, is produced as a by-product of nuclear fission in uranium fuel. It is also the main material used in nuclear weapons (including H-bombs, which use fission to provide the high temperatures needed for hydrogen fusion). But U-235, suitably concentrated ('enriched'), can also be used for weapons. So to make a nuclear bomb you need either an enrichment system to concentrate U-235 or a reactor to make plutonium of the right grade. Enrichment technologies used include diffusion baffles (a multiple filtering approach) and the more efficient centrifuge approach.

Most reactors (Canada's Candu plants and the older UK Magnox plants apart) need slightly enriched uranium to run. So close monitoring is required to determine whether a specific enrichment activity is being used to produce fuel for civil nuclear power or (with higher-level enrichment) to produce material for nuclear weapons. Similarly, it can be difficult to know whether reactors are being used to make military-grade plutonium. In other words, nuclear weapons and nuclear power are inextricably linked; some say they are two sides of the same coin.

Uranium mining, ore processing, fuel fabrication and enrichment are energy-intensive processes. Since at present most of the energy these processes use comes from fossil sources, the nuclear fuel cycle is not carbon-free. Indeed, as reserves of high-grade uranium ore are depleted, and lower-grade ores are used, the carbon debt will rise – unless non-fossil sources can be used to supply the fuel-processing energy. Ultimately, in theory, nuclear sources could be used to supply this power, but then we would run into the issue of the limited uranium resources being used up faster.

The main known reserves of uranium are in Australia, Canada, Namibia and Kazakhstan, and are said to be enough to keep existing plants going for around 80 years at current use rates. However, there could be a shortfall in the short term, since current use rates are around 69,000 tonnes a year whereas production is only around 54,000 tonnes (Energy Watch Group, 2006).

The gap is being made up, for the moment, by the use of material released from the closure of weapons programmes and the dismantling of nuclear weapons. However, this is a one-off option.

Longer term, new finds of uranium and new uranium-using technology may help extend the resource. For example, fast neutron breeder

DOI: 10.1057/9781137274335

reactors could help stretch uranium reserves by 'breeding' plutonium from otherwise wasted uranium.

Some prototypes have been built, but so far this is a relatively unde-veloped technology, with potentially significant safety and security problems. The spent fuel must be reprocessed to extract the plutonium, and then the wastes must be dealt with.

The US breeder programme was halted in the 1970s because of concerns about the risk of plutonium proliferation. Subsequently, the UK and France backed off from this option on account of the high costs and uncertain eco-nomics, while Japan's fast breeder prototype was closed down after a sodium fire. (Liquid sodium is used as the coolant in these plants; Cochran, 2010.)

Other new types of nuclear technology might address some of the safety, cost, proliferation, waste and fuel depletion issues. The EPR and AP1000 are examples of 'third-generation' technologies, basically evolutionary develop-ments of the PWR. But a range of fourth-generation options is emerging.

One version makes use of thorium. Thorium is attractive because uranium reserves are relatively limited and there is about three times as much thorium in the world as uranium. But it has the disadvantage that it is not fissile. To make a nuclear reaction you have to supply neutrons, either from a particle accelerator or from conventional nuclear fission using uranium or plutonium to produce a new material, U-233, which is fissile. Both approaches are being followed up, for example with sub-critical thorium reactor designs using accelerators and with molten-salt fluoride-based systems using thorium and plutonium (http://www.the-weinberg-foundation.org/).

It has been claimed that thorium-based systems would produce less waste and that, with some types of breeder system, it would also be possi-ble to burn up some wastes. Although China and India, among others, are exploring these ideas, we are still some way (probably a decade or more) from commercial-scale reactors, and there are many unknowns, includ-ing, crucially, the cost and safety (http://www.nnl.co.uk/positionpapers).

Despite decades of research and $20 billion or more in funding, we are even further away from developing commercial-scale nuclear fusion, sometimes seen as the 'holy grail' of energy technology, fusing light ele-ments together at very high temperatures to release energy, as happens in the sun (and in H-bombs).

Another $20 billion or so might see some useful results from either the US laser-fired 'implosion' compression system or ITER, the inter-national 'magnetic constriction' project being built in France. But we would still be a long way from commercial power reactors, and there

DOI: 10.1057/9781137274335

would still also be many uncertainties about safety and costs. The UK Atomic Energy Authority has suggested that fusion 'has the potential to supply 20% of the world's electricity by the year 2100'.

Renewables already supply about 20% now, and supporters sometimes ask why we are spending billions on long shots like this, which at best are unlikely to deliver significant power for many decades, when we already have a working fusion reactor in the sky, the sun, delivering more energy than we could ever use if we can tap it effectively and efficiently.

A.3 Radiation units and impacts

In the International System of units (SI), the becquerel (Bq) is the unit of radioactivity. One becquerel is defined as one nuclear disintegration per second. It's a relatively small unit, so multiples are common – millions, billions or trillions of becquerels, the latter being designated TBq, for terabecquerel (i.e. one thousand billion becquerels). It has replaced an earlier unit, the curie, which was the activity of one gram of pure radium (1 curie = 3.7 billion Bq).

Radioactive material does not stay active at the same level indefinitely. In a fixed mass of radioactive material, the activity level falls with time. The time taken for the activity level of a fixed amount of radioactive material to fall by half (more strictly, for half the nuclear decays in a sample to have occurred) is known as the half-life. For example, the half-life of plutonium-239 is around 24,000 years.

Radiation doses are often measured in sieverts (Sv), or more usually, since this is a large unit, millisieverts (mSv; 1,000 mSv = 1 Sv). It is actually a dose-*equivalent* radiation measure, attempting to quantitatively add together the biological effects of any type of ionising radiation, compared to an equal dose of gamma rays. The effects of radiation exposure are cumulative, but they are usually assumed to be worse if the dose is received rapidly, so the rate of exposure matters as well as the overall amount. Nevertheless, exposure limit levels are usually set in terms of, for example, total annual doses or (very occasionally) lifetime doses.

A radiation dose of 500 mSv or more can cause acute symptoms within a few days. Studies of those exposed to radiation from the atomic bomb blast at Hiroshima showed that about 50% of those who received a whole-body dose of 4,500 mSv died. The level set for relocation from the Chernobyl area was a dose of 350 mSv in a lifetime.

DOI: 10.1057/9781137274335

Typically, depending on where you live, you are exposed to about 2–3 mSv each year of background radiation from cosmic rays, soil radionuclides, radon gas and other sources. However, background radiation is not harmless: it can play a role in causing cancers and other health effects.

The public limit for extra non-natural radiation exposure for adults in most Western countries is 1 mSv per year. Occupational exposure limits are set higher than those for the general public, in part since it is argued that levels and exposures can be more carefully monitored and controlled. Typically, the limit is now set at around 20 mSv per annum, although levels and standards vary around the world.

Some experts see these limits as too lax, given current and emerging knowledge about radiation biology, which has focused on new models of the interaction between radiation and cells, including the so-called bystander effect on cells not directly hit and genomic instability in later generations of cell clones remote from the radiation exposure. Depending on the radiation levels involved, there is also likely to be a large difference in impact between external exposure to radiation and exposure through ingestion, inhalation or absorption of particles, especially long term.

One key issue is the nature of the dose–response relationship. All of the world's official bodies (including the International Commission on Radiological Protection, the UN Scientific Committee on the Effects of Atomic Radiation, the Committee on the Biological Effects of Ionizing Radiation and the International Atomic Energy Agency) assume that the relationship is linear all the way down to zero dose; that is, small risks exist no matter how low the dose. However, some scientists say that radiation damage to cells does not fall off linearly with reduced dose, but can be lower, and others argue the opposite, making the point that particles ingested, inhaled or absorbed internally (e.g. into the gut or lung) are likely to have much more impact over time than short external exposure to radiation. So, they claim, the comparisons often made by nuclear apologists between low external exposure levels from nuclear leaks or accidents and background radiation levels are unhelpful and, indeed, irrelevant.

The crux of the matter is that current official dose estimates from internal exposures are considered to contain large uncertainties (http://www.cerrie.org). In other words, the risks from internal exposures (e.g. from the ingestion or inhalation of nuclides released by nuclear facilities) could be much larger than currently estimated (http://www.safegrounds.com/radiation_risk.htm).

DOI: 10.1057/9781137274335

There is also the issue of concentration as a radioactive material passes through the food chain. This depends on its chemical form and which organisms take it up. A 1999 study found that seals and porpoises in the Irish Sea concentrated radioactive caesium by a factor of 300 relative to its concentration in seawater, and by a factor of 3–4 compared with the fish they ate (Grossman, 2011; Schiermeier, 2011).

A.4 Fukushima doses

The level set by the Japanese authorities for nuclear workers at Fukushima was raised initially from 50 mSv to 100 mSv per annum, but some workers exceeded this higher level during their efforts to deal with the crisis. For example, two workers received doses of 170–180 mSv on 24 March 2011. A revised level of 250 mSv per annum was introduced, and it seems several workers exceeded even that. The limit was later reduced back to 100 mSv, still five times the accepted level in most other countries.

There were reports from Japan's Health and Labour Ministry that, by June 2011, 8 workers had exceeded the 250 mSv limit and at least 90 others had exceeded the earlier limit of 100 mSv, including several who were nearing the higher limit. Two control room operators had been exposed to more than 600 mSv. There were reports that at one stage levels reached 400 mSv *per hour* and above around the site.

For further information on radiation and its effects, see Warry (2011), Areva (2011) and Rogers (2011).

DOI: 10.1057/9781137274335

Credits and Afterword

Parts of the text in Chapters 1 and 2 and in the Appendix draw on material I used for an article for SciDev.net (Elliott, 2011b). Some other parts come from an article I produced for the Scientists for Global Responsibility *Newsletter* (issue 40; http://www.sgr.org.uk) and from *Renew*, the newsletter I edit (http://www.natta-renew.org).

Thanks are due to Tam Dougan for her useful inputs on food chain issues and other matters, to Oliver Elliott for his critical advice and to Dr Ian Fairlie for his helpful inputs, including trying to guide me through the labyrinthine debate over radiation biology and nuclear safety. I fear that, as you will see from his work at http://www.ianfairlie.org, I have only scratched the surface.

This is the third book I have written or edited on nuclear issues. I have to say that I would have preferred instead to put my efforts into more positive studies of issues surrounding the development of renewable energy. But it seems, as ever, and arguably unfairly, that the nuclear issue still often dominates the energy scene.

In surveying, and then analysing, the reactions to Fukushima, I have had to rely partly on media reports, some of which may be uncorroborated. Some may also reflect partisan anti- or pro-nuclear views. I have sought to provide a balance of views and to avoid getting too drawn into the pro/anti debate, given that I have explored that elsewhere – although my own stance will perhaps inevitably be clear. In common with most environmentalists, I see the continued use of fossil fuels, coal especially, as a major problem, because of climate and air pollution

DOI: 10.1057/9781137274335

issues, but I am not convinced that nuclear provides a viable alternative, whereas renewables could well do. After Fukushima it could be that this view will become more popular.

I handed the final version of this text over for publication not long after the first anniversary of the Fukushima disaster. Since then, two reactors have been restarted, despite major protests. In parallel, the report of the Fukushima Nuclear Accident Independent Investigation Committee came out, concluding that the accident was 'manmade', the result of wilful negligence and flawed responses (http://naiic.go.jp/en/). In addition, a study emerged from Stanford University suggesting that radiation-related mortality and morbidity could be of the order of hundreds or perhaps thousands (Ten Hoeve and Jacobson, 2012). I would like to dedicate this book to all those who suffered, and continue to suffer, as a result of this tragic event and its aftermath. For many of them, a current focus will be the government consultation on whether to aim for zero, 15% or 20–25% nuclear by 2030. In September 2012, in something of a compromise, the Japanese Cabinet indicated that it would aim to get to zero nuclear in the 2030s.

If you want to follow up on the renewable energy options and related policies, see my 'Renew On Line' bimonthly newsletter (http://www.natta-renew.org), my monthly overview (http://delliott6.blogspot.com) and my weekly 'Renew your Energy' blog (http://environmentalresearchweb.org/blog/renew-your-energy).

DOI: 10.1057/9781137274335

References

Full web links have been provided where available. These sites were all accessed in March 2012.

ABJ (2011) 'NRG Energy Halts S. Texas Nuclear Plant Expansion', *Austin Business Journal*, Texas, 20 April. http://www.bizjournals.com/austin/news/2011/04/20/nrg-energy-stops-funding-s-texas.html

AIE/AEA (2011) 'Energy Balance of Nuclear Power Generation', Austrian Institute of Ecology/Austrian Energy Agency, Vienna. http://www.ecology.at/lca_nuklearindustrie.htm

Aniletto (2011) 'Is Malaysia's Nuclear Power Programme Really On Hold?' Aniletto website news, 31 October. http://anilnetto.com/economy/energy-resources/are-malaysia-s-nuclear-power-plans-really-on-hold/

ARCH (2012) International Agency for Research on Cancer, Lyon (part of the World Health Organization) website. http://www.iarc.fr/

Areva (2011) 'Radiation Dose Chart Makes Sense of Sieverts', 22 March. http://us.arevablog.com/2011/03/22/radiation-dose-chart-makes-sense-of-sieverts/

ASI (2011) 'Renewable Energy: Vision or Mirage', Adam Smith Institute, London.

Bailey, R., and Blair, L. (2012) 'A Corruption of Governance', Unlock Democracy and the Association for the Conservation of Energy, London.

Balonov, M. (2011) review of *Chernobyl: Consequences of the Catastrophe for People and the Environment* by A. Yablokov *et al.* (2009), *Annals of the New York Academy of Sciences*, September. http://www.nyas.org/

DOI: 10.1057/9781137274335

publications/annals/Detail.aspx?cid=f3f3bd16-51ba-4d7b-a086-753f44b3bfc1

Banks, J., Ebinger, C., Massy, K., and Avasarala, G. (2012) 'Models for Aspirant Civil Nuclear Energy Nations in the Middle East', The Brookings Institution, Washington DC, 12 February. http://www.brookings.edu/papers/2011/0927_middle_east_nuclear_ebinger_banks.aspx

BBC (2012) BBC website statement. http://www.bbc.co.uk/complaints/comp-reports/ecu/banggoesthetheory

BBC World Service (2011) 'Opposition to Nuclear Energy Grows: Global Poll', BBC GlobeScan survey. http://www.globescan.com/news_archives/bbc2011_energy/

Becker, M. (2011) 'Japanese Nuclear Plant Operator Plagued by Scandal', Spiegel Online, Berlin, 23 March. http://www.spiegel.de/international/world/0,1518,752704,00.html#ref=nlint

Berger, A. (2011) 'Renaissance On Hold: Public Opinion and Nuclear Energy in Southeast Asia', undated web commentary, Royal United Services Institute, London. http://www.rusi.org/analysis/commentary/ref:C4E7C527C416AA/

BERR (2007) 'Energy White Paper: Meeting the Energy Challenge', Department of Business, Enterprise and Regulatory Reform, London.

Bethge, P., and Knauer, S. (2006) 'How Close Did Sweden Come to Disaster?' Spiegel Online, Berlin, 7 August. http://www.spiegel.de/international/spiegel/0,1518,430458,00.html

Beyond Nuclear (2011) 'Nuclear Power Banned in Italy', grassroots group web coverage. http://www.beyondnuclear.org/home/2011/6/13/victory-its-official-nuclear-power-banned-in-italy.html

Bird, W. (2012) 'Fukushima Nuclear Cleanup Could Create Its Own Environmental Disaster', *Guardian*, London, 9 January. http://www.guardian.co.uk/environment/2012/jan/09/fukushima-cleanup-environmental-disaster?INTCMP=SRCH

Black, R. (2011) 'Nuclear Power "Gets Little Public Support Worldwide"', BBC web news item on a GlobeScan poll for the BBC, 25 November. http://www.bbc.co.uk/news/science-environment-15864806

Blair, T. (2006) 'Blair: Nuclear Back on the Agenda', speech to the CBI, 16 May. http://www.politics.co.uk/news/2006/5/17/blair-nuclear-back-on-the-agenda

Bloomberg (2011) 'Sun Sets on Oil for Gulf Power Generation', white paper, Bloomberg New Energy Finance, 19 January. http://www.bnef.com/free-publications/white-papers/

DOI: 10.1057/9781137274335

Blowers, A. (2012) 'Fukushima – It Is a Moral Issue', *Town and Country Planning Journal*, Town and Country Planning Association, London, March, pp. 141–149.

BMU (2011a). 'Development of Renewable Energy Sources in Germany 2010', BMU, German Federal Ministry for the Environment, Nature Conservation and Nuclear Safety. http://www.bmu.de/files/english/pdf/application/pdf/ee_in_deutschland_graf_tab_en.pdf

BMU (2011b) 'The Federal Government's Energy Concept of 2010 and the Transformation of the Energy System of 2011', BMU, German Federal Ministry for the Environment, Nature Conservation and Nuclear Safety. http://www.bmu.de/files/english/pdf/application/pdf/energiekonzept_bundesregierung_en.pdf

Boyle, G. (ed.) (2007) *Renewable Electricity and the Grid: The Challenge of Variability*, Earthscan, London.

BP (2012) 'Energy Outlook 2030', British Petroleum report. http://www.bp.com/energyoutlook2030

Brenhouse, H. (2011) 'Fukushima Blast: Are Japan's Neighbors At Risk?' *Time* newsfeed, 15 March. http://newsfeed.time.com/2011/03/15/fukushima-blast-are-japans-neighbours-at-risk/

Brenner, D., Doll, R., Goodhead, D., Hall, E., Land, C., Little, J., Lubing, J., Preston, D., Preston, J., Puskin, J., Ron, E., Sachs, R., Samet, J., Setlow, R., and Zaider, M. (2003) 'Cancer Risks Attributable to Low Doses of Ionizing Radiation: Assessing What We Really Know', *PNAS* 100(24), 25 November, pp. 13761–13766.

Brumfiel, G. (2011) 'Fallout Forensics Hike Radiation Toll: Global Data on Fukushima Challenge Japanese Estimates', *Nature* 478, 25 October, pp. 435–436. http://www.nature.com/news/2011/111025/full/478435a.html

BSA (2011) 'Nuclear Fallout', British Science Association website. http://www.britishscienceassociation.org/web/News/FestivalNews/nuclearpoll.htm

Burke, T. (2012) 'Nucléaire nouvelle génération? Non merci!' ENDS Report 445, February.

Burke, T., Juniper, T., Porritt, J., and Secrett, C. (2012) 'Investing in Nuclear Power: Current Concerns', Briefing No. 2, for the government, 4 April.

Busby, C. (2011) 'The Health Outcome of the Fukushima Catastrophe: Initial Analysis from Risk Model of the European Committee on Radiation Risk', European Committee on Radiation Risk, March. http://llrc.org/fukushima/subtopic/fukushimariskcalc.pdf

DOI: 10.1057/9781137274335

Busby, J. (2011) 'EdF's Financial Meltdown', After Oil website. http://
www.after-oil.co.uk/edf_financial.htm

Butler, D. (2011) 'Future of Chernobyl Health Studies in Doubt', *Nature*,
30 September. http://www.nature.com/news/2011/110930/full/
news.2011.565.html

Caracappa, P. (2011) 'Fukushima Accident: Radioactive Releases and
Potential Dose Consequences', paper presented at the ANS Annual
Meeting, June. http://www.ans.org/misc/FukushimaSpecialSession-
Caracappa.pdf

Carbon Trust (2011) 'Accelerating Marine Energy' Carbon Trust Report
CTC797, Carbon Trust, London, July. http://www.carbontrust.co.uk/
news/news/press-centre/2011/Pages/MEA.aspx

Carpenter, S. (2012) *Japan's Nuclear Crisis: The Routes to Responsibility*,
Palgrave Macmillan, Basingstoke.

CAT (2010) 'Zero Carbon Britain', Centre for Alternative Technology,
Machynlleth. http://www.zerocarbonbritain.com/

CERRIE (2004) 'Report of the Committee Examining Radiation Risks
of Internal Emitters', Committee Examining Radiation Risks of
Internal Emitters, London.

Charles, M. (2010) review of *Chernobyl: Consequences of the Catastrophe
for People and the Environment* by A. Yablokov *et al.* (2009), in
Radiation Protection Dosimetry 141(1), pp. 101–104.

Chillymanjaro (2011) 'The Waste from the Japanese Earthquake and
Tsunami', The Watchers website, 29 September. http://thewatchers.
adorraeli.com/2011/09/29/the-waste-from-the-japanese-earthquake-
and-tsunami/

China Post (2011) 'Taiwan Still Safe in Worst-Case Fukushima Scenario:
AEC', *China Post* report, 17 March. http://www.chinapost.com.tw/
asia/japan/2011/03/17/294988/Taiwan-still.htm

CIVITAS (2012) 'Electricity Costs: The Folly of Windpower', CIVITAS,
London.

CNIC (2011) Citizens' Nuclear Information Centre, Tokyo, website.
http://www.cnic.jp/english

CNTV (2011) 'Chinese Sea Possibly Contaminated by Radiation from
Fukushima', China Network Television website, 16 June. http://
english.cntv.cn/program/china24/20110816/101974.shtml

Cochran, T. *et al.* (2010) 'Fast Breeder Reactor Programs: History
and Status', a research report of the International Panel on Fissile
Materials, February. http://fissilematerials.org/library/rr08.pdf

DOI: 10.1057/9781137274335

COMARE (2011) 'Further Consideration of the Incidence of Childhood Leukaemia around Nuclear Power Plants in Great Britain', 14th Report of the Committee on Medical Aspects of Radiation in the Environment, London. http://www.comare.org.uk/press_releases/14thReportPressRelease.htm

Cook, T., and Elliott, D. (2012) 'New Europe, New Energy', in S. Shmelev *et al.* (eds), *Sustainability Analysis: An Interdisciplinary Approach*, Palgrave Macmillan, Basingstoke.

Cooper, M. (2009) 'The Economics of Nuclear Reactors: Renaissance or Relapse?' Institute for Energy and the Environment, Vermont Law School, June. Summary at http://www.nirs.org/mononline/nm692_3.pdf

Cooper, M. (2012) 'Nuclear Safety and Nuclear Economics: Fukushima Reignites the Never-Ending Debate', Symposium on the Future of Nuclear Power, University of Pittsburgh, 27–28 March. http://www.vermontlaw.edu/energy/news

Crooks, E. (2011) 'Nuclear: Enthusiasm for Reactor Investment Cools', *Financial Times*, London, 28 September. http://www.ft.com/cms/s/0/765c917c-e836-11e0-9fc7-00144feab49a.html#axzz1smC4q5JM

Davis, L. (2011) 'Prospects for U.S. Nuclear Power after Fukushima', The Energy Institute at Haas, Working Paper 218, University of California, Berkeley, August. http://ei.haas.berkeley.edu/pdf/working_papers/WP218.pdf

DECC (2010) Youth Panel report for the UK Department of Energy and Climate Change. http://www.decc.gov.uk/en/content/cms/news/pn10_121/pn10_121.aspx

Dennis, W. (2011) 'China May Double PV Plans after Nuclear Crisis', *Engineering and Technology Magazine*, IET, London, 19 April. http://eandt.theiet.org/news/2011/apr/china-pv.cfm

Dickie, M., and Cookson, C. (2011) 'Nuclear Energy: A Hotter Topic Than Ever', *Financial Times*, London, 11 November. http://www.ft.com/cms/s/0/aa0a40ec-0aea-11e1-b62f-00144feabdc0.html#ixzz1dJ6hyjqR

DOE (2011) *Annual Energy Outlook 2011*, Early Release Overview, US Department of Energy, Energy Information Administration, Washington, DC, February.

Dorfman, P. (2012) 'Nuclear Risk Assessment post-Fukushima Dai-ichi', *Energy and Environmental Risk Management*, 12 March. http://www.eaem.co.uk/opinions/nuclear-risk-assessment-post-fukushima-dai-ichi

DOI: 10.1057/9781137274335

Doro-chibra (2011) Doro-chibra: National Railway Motive Power Union of Chibra, International Workers' Solidarity Rally, Tokyo, November. http://www.doro-chiba.org/english/english.htm and local action at http://www.youtube.com/watch?v=yvxscUDKLXA

DTI (2003) 'Energy White Paper 2003: Our Energy Future – Creating a Low Carbon Economy', Department of Trade and Industry, London.

Dvorak, P. (2012) 'Japan Struggles with Tainted Reactor Water', *Wall Street Journal*, 29 February. http://online.wsj.com/article/SB10001424052970203833004577251150563609254.html?mod=WSJEUROPE_hpp_MIDDLEThirdNews

EC (2009) 'EmployRES', report for the European Commission by the Fraunhofer Institute, Ecofys, EEG and others, April. http://ec.europa.eu/energy/renewables/studies/doc/renewables/2009_employ_res_report.pdf

ECF (2010) 'Roadmap 2050', European Climate Foundation, Brussels. http://www.roadmap2050.eu

EDF (2008) 'EDF Energy Submission to the Department for Business, Enterprise & Regulatory Reform/Department of Energy and Climate Change Consultation on UK Renewable Energy Strategy', Log Number 00439e p.3. http://decc.gov.uk/en/content/cms/consultations/cons_res/cons_res.aspx

Edwards, R. (2011) 'Revealed: British Government's Plan to Play Down Fukushima', *Guardian*, London, 30 June. http://www.guardian.co.uk/environment/2011/jun/30/british-government-plan-play-down-fukushima

Elliott, D. (1988) 'Nuclear Power and the UK Trade Union and Labour Movement', Technology Policy Group, Occasional Paper 17, The Open University, Milton Keynes, October.

Elliott, D. (ed.) (2010) *Nuclear or Not?* Palgrave Macmillan, Basingstoke.

Elliott, D. (2011a) 'Green Heat – District Heating and Energy Storage', Environmental Research Web, Renew your energy. http://environmentalresearchweb.org/blog/2011/01/green-heat---district-heating.html

Elliott, D. (2011b) 'Nuclear Power after Fukushima: Facts and Figures', SciDev.net, 28 September. http://www.scidev.net/en/climate-change-and-energy/nuclear-power-after-fukushima/features/nuclear-power-after-fukushima-facts-and-figures-1.html

Elliott, D., and Cook, T. (2004) 'Symbolic Power: The Future of Nuclear Energy in Lithuania', *Science as Culture* 13(3), September, pp. 373–400.

DOI: 10.1057/9781137274335

ENE News (2011) 'Leaked TEPCO Report: 120 Billion Becquerels of Plutonium, 7.6 Trillion Becquerels of Neptunium Released in First 100 Hours – Media Concealed Risk to Public', ENE News, Tokyo, 15 October. http://enenews.com/leaked-tepco-report-120-billion-becquerels-of-plutonium-7-6-trillion-becquerels-of-neptunium-released-in-first-100-hours-media-concealed-risk-to-public

ENE News (2012) '7 Reports of Nuclear Fuel Rod Pieces Being Ejected from Fukushima Reactors and/or Spent Fuel Pools', ENE News, Tokyo, 25 February. http://enenews.com/pieces-nuclear-fuel-ejected

Energy Watch Group (2006) 'Uranium Resources and Nuclear Energy', December. http://www.energywatchgroup.org/fileadmin/global/pdf/EWG_Report_Uranium_3-12-2006ms.pdf

EREC (2010) 'Rethinking 2050', European Renewable Energy Council, Brussels. http://www.rethinking2050.eu

ERP (2012) 'UK Nuclear Fission Technology Roadmap', produced for the Energy Research Roadmap by the UK National Nuclear Laboratory, February. http://www.energyresearchpartnership.org.uk/nucleartechnologyroadmap

Ethics Commission (2011) 'Germany's Energy Turnaround: A Collective Effort for the Future', Ethics Commission on a Safe Energy Supply, report to the German Federal government, May.

EWEA (2012) 'French Nuclear Set to Become More Expensive Than Wind', European Wind Energy Association, February. http://blog.ewea.org/2012/02/french-nuclear-set-to-become-more-expensive-than-wind-power/

Exxon (2012) 'Energy Outlook', Exxon Mobil. http://www.exxonmobil.com/corporate/energy_outlook_eoelectricitycost.aspx

Fackler, M. (2012) 'Japanese Struggle to Protect Their Food Supply', *New York Times*, 21 January. http://www.nytimes.com/2012/01/22/world/asia/wary-japanese-take-food-safety-into-their-own-hands.html?_r=1

Fairlie, I. (2009) 'Commentary: Childhood Cancer Near Nuclear Power Stations', *Environmental Health* 8. http://www.ehjournal.net/content/8/1/43

Fairlie, I. (2010) 'The Risks of Nuclear Energy Are Not Exaggerated', *Guardian*, London, 20 January. http://www.guardian.co.uk/commentisfree/2010/jan/20/evidence-nuclear-risks-not-overrated

Fairlie, I., and Sumners, D. (2006) 'TORCH: The Other Report on Chernobyl', produced for the Greens/EFA in the European Parliament. http://www.chernobylreport.org

DOI: 10.1057/9781137274335

Ferguson, C., and Settle, F. (eds.) (2012) 'The Future of Nuclear Power in the United States', Federation of American Scientists. http://www.fas. org/pubs/_docs/Nuclear_Energy_Report-lowres.pdf

Fox, F. (2011) 'Media Meltdown over Nuclear Threat', Science Media Centre, London, as relayed by the BBC College of Journalism website, 25 March. http://www.bbc.co.uk/journalism/blog/2011/03/ british-media-went-into-meltdo.shtml

Froggatt, A. (2010) 'Systems for Change: Nuclear Power vs. Energy Efficiency + Renewables?' paper prepared with Mycle Schneider for the Heinrich Böll Foundation, Berlin, March. http://www.boell.de

Funabashi, Y. (2012) report from the Independent Investigation Commission on the Fukushima Daiichi Nuclear Accident, a committee of the Rebuild Japan Initiative Foundation, led by Yoichi Funabashi, as covered in an Asahi Shimbun news report/interview, 29 February. http://ajw.asahi.com/article/0311disaster/fukushima/ AJ201202290078

Funabashi, Y., and Kitatazawa, K. (2012) 'Fukushima in Review: A Complex Disaster, a Disastrous Response', *Bulletin of the Atomic Scientists*, March. http://bos.sagepub.com/content/early/2012/02/29/0 096340212440359

Gamson, W., and Modigliani, A. (1989) 'Media Discourse and Public Opinion on Nuclear Power; A Constructionist Approach', *American Journal of Sociology* 95(1), pp. 1–37. http://www.jstor.org/pss/2780405

German Federal Environment Agency (2011) 'Restructuring Electricity Supply in Germany', background paper, Umwelt Bundes Amt, May. http://www.umweltbundesamt.de

Goldemberg, J. (2009) 'Nuclear Energy in Developing Countries', *Daedalus*, the American Academy of Arts & Sciences, 138(4), pp. 71–80.

Goldemberg, J. (2011) 'Have Rising Costs and Increased Risks Made Nuclear Energy a Poor Choice?' 'Oil Price' blog, 4 October. http:// oilprice.com/Alternative-Energy/Nuclear-Power/Have-Rising-Costs- and-Increased-Risks-Made-Nuclear-Energy-a-Poor-Choice.html

Green Facts (2012) 'Chernobyl Nuclear Accident', fact file. http://www. greenfacts.org/en/chernobyl/

Greenpeace (2003) 'Energy Rich Japan', Greenpeace Germany report. http://www.energyrichjapan.info

Greenpeace (2011) 'The Advanced Energy [R]evolution: A Sustainable Energy Outlook for Japan', Greenpeace/EREC report. http://www. greenpeace.org/japan/Global/japan/pdf/er_report.pdf

DOI: 10.1057/9781137274335

Greenpeace (2012) 'The Lessons from Fukushima', Greenpeace report and press release, February. http://www.greenpeace.org/international/en/news/Blogs/nuclear-reaction/less=ons-from-fukushima-new-greenpeace-report-/blog/39271/

Grimes, R., and Nuttall, W. (2010) 'Generating the Option of a Two-Stage Nuclear Renaissance', *Science*, 13 August, pp. 799–803.

Grossman, E. (2011) 'Radioactivity in the Ocean: Diluted, but Far from Harmless', environment360. http://e360.yale.edu/feature/radioactivity_in_the_ocean_diluted_but_far_from_harmless/2391/

Guardian (2012) 'A Power Station in Your Back Yard?' *Guardian*/ICM 2012 poll; Ipsos MORI/Cardiff University 2010 data, 1 March. http://www.guardian.co.uk/environment/2012/mar/01/local-opposition-onshore-windfarms-tripled?

Harper, P. (2011) 'Zero Carbon Japan', *Bio City* 48 (in Japanese), pp. 64–68.

Hiroshima City and Nagasaki City (1981) 'Hiroshima and Nagasaki: The Physical, Medical, and Social Effects of the Atomic Bombings', edited by the Committee for the Compilation of Materials on Damage Caused by the Atomic Bombs in Hiroshima and Nagasaki. English translation. Originally published in Japanese by Iwanami Shoten Publishers, Tokyo, 1979.

Hoedt, R. (2012) 'Japan's "Nuclear Village": Too Big to Fail?' *European Energy Review*, 6 February. http://www.europeanenergyreview.eu/site/pagina.php?id=3505

Höglund, L. (2006) speech to the IPPNW World Congress in Helsinki, Finland Plenary 'Sustainable Energy through Sustainable Security', 10 September. http://www.ippnw-europe.org/commonFiles/pdfs/Verein/Speech_Lars_Hoeglund.pdf

HSE (2011) UK Health and Safety Executive, Office for Nuclear Regulation, Weightman reports and EU Stress Test reports. http://www.hse.gov.uk/nuclear/

IAEA (2006) 'Chernobyl's Legacy: Health, Environmental and Socio-Economic Impacts and Recommendations to the Governments of Belarus, the Russian Federation and Ukraine', The Chernobyl Forum: 2003–2005, International Atomic Energy Agency, Vienna. http://www.unscear.org/unscear/en/chernobyl.html

IEER (2006) 'France Can Phase Out Nuclear Power and Achieve Low Carbon Dioxide Emissions', Institute for Energy and Environmental Research, Maryland. http://www.ieer.org/reports/energy/france/

IISS (2011) 'The Fallout from Fukushima', strategic comments, International Institute for Strategic Studies, London, May. http://www.iiss.org/publications/strategic-comments/past-issues/volume-17-2011/may/the-fallout-from-fukushima/

INPO (2011) 'Special Report on the Nuclear Accident at the Fukushima Daiichi Nuclear Power Station', Institute of Nuclear Power Operations, INPO 11-005, November.

IPCC (2011) 'Special Report on Renewable Energy Sources and Climate Change Mitigation (SRREN)', Intergovernmental Panel on Climate Change, Geneva. http://www.srren.org

Ipsos (2011a) 'Global Citizen Reaction to the Fukushima Nuclear Plant Disaster', Ipsos Global Advisor, global poll carried out in May, published in June. http://www.ipsos-mori.com/Assets/Docs/Polls/ipsos-global-advisor-nuclear-power-june-2011.pdf

Ipsos (2011b) 'Public Attitudes to the Nuclear Industry', UK poll for the Nuclear Industry Association in June, Ipsos MORI. http://www.ipsos-mori.com/researchpublications/researcharchive/2834/Public-attitudes-to-the-nuclear-to-the-nuclear-industry.aspx

Ipsos (2012) 'After Fukushima: Global Opinion on Energy Policy', Ipsos Social Research Institute, March. http://www.ipsos.com/public-affairs/sites/www.ipsos.com.public-affairs/files/Energy%20Article.pdf

IRSN (2011) coverage of French Institut de Radioprotection et de Sûreté Nucléaire, Paris, report in English, 28 October. http://www.maritime-executive.com/article/fukushima-disaster-produces-world-s-worst-nuclear-sea-pollution

IRSN (2012) '2012 Barometer: IRSN Perception of Risks and Safety for the French', Institut de Radioprotection et de Sûreté Nucléaire, Paris.

Ishizuka, H., and Mori, H. (2011) 'Estimated 13,000 Square km Eligible for Decontamination', Asahi news service, 12 October. http://www.asahi.com/english/TKY201110110214.html

IWES (2010) 'Renewable Energies and Base Load Power Plants: Are They Compatible?' Fraunhofer Institute for Wind Energy and Energy System Technology, Berlin. http://www.unendlich-viel-energie.de/en/details/article/523/renewable-energies-and-base-load-power-plants-are-they-compatible.html

Jacobson, M., and Delucchi, M. (2009) 'A Path to Sustainable Energy by 2030', *Scientific American*, November. http://www.stanford.edu/group/efmh/jacobson/Articles/I/susenergy2030.html

DOI: 10.1057/9781137274335

Jacobson, M., and Delucchi, M. (2011) 'Providing All Global Energy with Wind, Water, and Solar Power', *Energy Policy* 39(3), March, pp. 1154–1190.

Jaczko, G. (2012) 'Looking to the Future', speech by the chair of the US Nuclear Regulatory Commission, Platts 8th Annual Nuclear Energy Conference, Rockville, MD, 9 February. http://www.neimagazine. com/journals/Power/NEI/March_2012/attachments/s-12-002.pdf

Japan Focus (2011) 'Japan's Irradiated Beef Scandal', *Asia-Pacific Journal* 9(30:5), 25 July. http://japanfocus.org/-Asia_Pacific_Journal-Feature/3577

Jargin, S. (2010) 'Overestimation of Chernobyl Consequences: Poorly Substantiated Information Published', letter to the editor, *Radiation and Environmental Biophysics* 49(4), pp. 743–745. http://www. springerlink.com/content/e706705592415435/

Jewell, J. (2011) 'A Nuclear-Powered North Africa: Just a Desert Mirage or Is There Something on the Horizon?' *Energy Policy* 39(8), August, pp. 4445–4457.

JWPA (2010) 'Long-Term Installation Goal on Wind Power Generation and Roadmap V2.1', Japan Wind Power Association. http://jwpa.jp/ page_132_englishsite/jwpa/detail_e.html

Kan, N. (2011a) Prime minister's speech at a press conference on 13 July. http://kantei.go.jp/foreign/kan/statement/201107/13kaiken_e.html

Kan, N. (2011b) 'Japan to Reconsider Energy Policy', World Nuclear News report of prime minister's speech, 11 May. http://www.world-nuclear-news.org/NP-Japan_to_reconsider_energy_policy-1105114. html

Kanady, S. (2012) 'Qatar's N-Power Plan "Economically Feasible"', Professional Reactor Operator Society website, 10 February. http:// www.nucpros.com/content/qatar%E2%80%99s-n-power-plan-%E2%80%98economically-feasible%E2%80%99

Kantei (2011) 'Report of Japanese Government to the IAEA Ministerial Conference on Nuclear Safety – The Accident at TEPCO's Fukushima Nuclear Power Stations', prime minister of Japan and his cabinet. http://www.kantei.go.jp/foreign/kan/topics/201106/ iaea_houkokusho_e.html

Kasturi, C. (2011) 'Fukushima Radiation Can't Reach India', *Hindustan Times*, 16 March. http://www.hindustantimes.com/India-news/ NewDelhi/Fukushima-radiation-can-t-reach-India/Article1-674149. aspx

DOI: 10.1057/9781137274335

Kidd, S. (2010) 'Energy to 2050 – Can It Fit In With Environmental Objectives?' *Nuclear Engineering International*, 21 December. http://www.neimagazine.com/story.asp?storyCode=2058500

Kidd, S. (2011) 'Nuclear in East Asia – the Hotbed?' *Nuclear Engineering International*, 6 December. http://www.neimagazine.com story.asp?sectioncode=147&storyCode=2061333

Kidd, S. (2012a) 'Nuclear As the Last Resort – Why and How?' *Nuclear Engineering International*, 19 January. http://www.neimagazine.com/story.asp?storyCode=2061613

Kidd, S. (2012b) 'Public Acceptance – Is It Causing Nuclear Cost Escalation?' *Nuclear Engineering International*, 20 April. http://www.neimagazine.com/story.asp?storyCode=2062210

Kitschelt, H. (1986) 'Political Opportunity Structures and Political Protest: Anti-Nuclear Movements in Four Democracies', *British Journal of Political Science* 16(1), January, pp. 57–85.

Kojo, M., and Litmanen, T. (2009) *The Renewal of Nuclear Power in Finland*, Palgrave Macmillan, Basingstoke.

Koopmans, R., and Duyvendak, J. (1995) 'The Political Construction of the Nuclear Energy Issue and Its Impact on the Mobilisation of Anti-Nuclear Movements in Western Europe', *Social Problems* 42(2), May, pp. 235–251.

Kuwait Times (2011) 'Kuwait No Longer Interested in Pursuing Nuclear Energy', *Kuwait Times*, 12 July. http://www.kuwaittimes.net/read_news.php?newsid=NzI3MzUoODM5OQ==

Lahn, G., and Stephens, S. (2011) 'Burning Oil to Keep Cool: The Hidden Energy Crisis in Saudi Arabia', Chatham House, London.

Large, J. (2011) 'Incidents, Developing Situation and Possible Eventual Outcome at the Fukushima Dai-ichi Nuclear Power Plants', Large Associates, London. http://www.largeassociates.com/

Large, J. (2012) Large Associates website, reports and presentations. http://www.largeassociates.com/

Lewis, L. (2011) 'In the Fierce Heat of Summer, It's Cool to Save Energy', *The Times*, London, 16 July.

Lovins, A. (2012) Preface to M. Schneider, A. Froggatt and S. Thomas, *World Nuclear Industry Status Report 2010–2011*, Worldwatch Institute, Washington DC.

Mainichi (2011) report on opinion poll. http://mdn.mainichi.jp/mdnnews/news/20110920p2a00m0na017000c.html

DOI: 10.1057/9781137274335

Mangano, J., and Sherman, J. (2012) 'An Unexpected Mortality Increase in the United States Follows Arrival of the Radioactive Plume from Fukushima: Is There a Correlation?' *International Journal of Health Services* 42(1), pp. 47–64. http://www.radiation.org/reading/pubs/HS42_1F.pdf

Mattei, J., Vial, E., Rebour, V., Liemersdorf, H., and Turschmann, M. (2001) 'Generic Results and Conclusions of Re-Evaluating the Flooding Protection in French and German Nuclear Power Plants', IPSN/GRS, Eurosafe Forum. http://www.eurosafe-forum.org

Maue, G. (2012), presentation on the German energy programme to a Parliamentary Renewables and Sustainable Energy Group seminar in London, February. http://dl.dropbox.com/u/15973585/PRASEG/120208%20%20Energy%20Concept%20short.ppt

McMahon, J. (2011) 'Radiation Detected in Drinking Water in 13 More US Cities, Cesium-137 in Vermont Milk', *Forbes* news review, 4 September. http://www.forbes.com/sites/jeffmcmahon/2011/04/09/radiation-detected-in-drinking-water-in-13-more-us-cities-cesium-137-in-vermont-milk/

McNeill, D. (2011a) 'Revealed: Secret Evacuation Plan for Tokyo after Fukushima', *Independent*, London, 27 January. http://www.independent.co.uk/news/world/asia/revealed-secret-evacuation-plan-for-tokyo-after-fukushima-6295353.html

McNeill, D. (2011b) 'Blood Money – Fukushima Victims Bitter over Compensation', *Asia-Pacific Journal*, 25 October. http://japanfocus.org/events/view/116

McNeill, D., and Adelstein, J. (2011) 'The Explosive Truth behind Fukushima's Meltdown', *Independent*, London, 17 August. http://www.independent.co.uk/news/world/asia/the-explosive-truth-behind-fukushimas-meltdown-2338819.html

MIC (2011) press release from Ministry of Internal Affairs and Communication on 'false rumours', April. http://www.soumu.go.jp/menu_news/s-news/01kibano8_01000023.html

Miller, S., and Sagan, S. (2009) 'Nuclear Power without Nuclear Proliferation?' *Daedalus*, American Academy of Arts and Sciences, 138(4), pp. 7–18.

Mitchell, C. (2011) 'Nuclear Power Is the Reason for the New Energy Regulations', *Guardian*, London, 11 March. http://www.guardian.co.uk/environment/2011/mar/11/nuclear-power-reason-energy-regulations

DOI: 10.1057/9781137274335

Mitchell, C., Froggatt, A., and Managi, S. (2012) 'Japanese Energy Policy Stands at a Crossroads', *Guardian*, London, 3 May. http://www. guardian.co.uk/environment/2012/may/03/japan-nuclear-power-post-fukushima

Modern Power Systems (2012) 'Bahrain Scraps Nuclear Power Plan', Modern Power Systems, 26 February. http://www.modernpowersystems. com/story.asp?sectionCode=131&storyCode=2061836

Monbiot, G. (2011) 'Why Fukushima Made Me Stop Worrying and Love Nuclear Power', *Guardian*, London, 21 March. http://www.guardian. co.uk/commentisfree/2011/mar/21/pro-nuclear-japan-fukushima

Morgan, J., and Matthes, F. (2011) 'How Germany Plans to Succeed in a Nuclear Free, Low Carbon Economy', WRI Insights website, 28 July. http://insights.wri.org/open-climate-network/2011/07/how-germany-plans-succeed-nuclear-free-low-carbon-economy

Morton, O. (2012) 'The Dream That Failed', *Economist*, 10 March. http:// www.economist.com/node/21549098

Moss, K. (2011) comment attributed to Ken Moss, CEO of mO3 Power, the UK's largest solar developer, as reported in Renew 191, NATTA, Milton Keynes, May–June.

Motts (2011a) 'UK Electricity Generation Costs – Update', Mott MacDonald consultants report for the UK Department of Energy and Climate Change, June. http://www.decc.gov.uk/assets/decc/ statistics/projections/71-uk-electricity-generation-costs-update-.pdf

Motts (2011b) 'Costs of Low-Carbon Technologies', Mott MacDonald consultants report for the Committee on Climate Change, May. http://hmccc.s3.amazonaws.com/Renewables%20Review/MML%20 final%20report%20for%20CCC%209%20may%202011.pdf

Moyer, M. (2011) 'Researchers Trumpet Another Flawed Fukushima Death Study', *Scientific American* blog. http://blogs.scientificamerican. com/observations/2011/12/20/researchers-trumpet-another-flawed-fukushima-death-study/

NCG (2011) UK Nuclear Consult Group website, complaint concerning a BBC *Bang Goes the Theory* programme transmitted on 3 October, and BBC interim response archive. http://www.nuclearconsult.com/ information.php

NEI (2012a) 'Noda Calls for Reduction of Nuclear in Japan, Where Only Three Reactors Remain in Operation', *Nuclear Engineering International* newsletter, 30 January. http://www.neimagazine.com/ story.asp?storyCode=2061649

DOI: 10.1057/9781137274335

NEI (2012b) 'Perspective on Public Opinion' series, including November 2011 report, Nuclear Energy Institute, US. http://www.nei.org/ resourcesandstats/publicationsandmedia/newslettersandreports/ perspectivesonpublicopinion/

NEI Nuclear Notes (2011) 'Joseph Mangano Contradicts His Own Press Release on Fukushima Research', NEI Nuclear Notes, independent website. http://neinuclearnotes.blogspot.com/2011/12/joseph-mangano-contradicts-his-own.html

News on Japan (2011) 'Fukushima Cleanup Could Cost up to \$250 Billion', press coverage of the Japan Center for Economic Research estimate. http://newsonjapan.com/html/newsdesk/article/89987.php

Novinte (2012) 'Bulgaria Quits Belene Nuclear Power Plant Project', Novinte-com news report, 28 March. http://www.novinite.com/ view_news.php?id=137961

NPS (2011) National Policy Statements for Energy Infrastructure, EN-1, Department of Energy and Climate Change, London. http://www. decc.gov.uk/en/content/cms/meeting_energy/consents_planning/ nps_en_infra/nps_en_ifra.aspx

Nuclear Intelligence Reports (2012), 'Nuclear Energy Market to 2020', industry review. http://www.industryreview.com/Report. aspx?ID=Nuclear-Energy-Market-to-2020--Technological-Innovations-New-Safety-Measures-and-Uptake-in-Asia-Pacific-to-Shape-Future-Development&ReportType=Industry_Report&coreindustry=ALL&Title=Nuclear_Energy

Nuclear Monitor (2011) 'EPR Construction in China: Same Problems', *Nuclear Monitor*, World Information Service on Energy, Amsterdam, 735, 2 October, p. 2.

Nuclear Monitor (2012) 'Lessons from Fukushima', *Nuclear Monitor*, World Information Service on Energy, Amsterdam, 743, 5 March.

Nuclear News (2012) news report, 19 January. http://nuclear-news.net/ category/2-world/middle-east/egypt/

Nuclear Review (1995) 'The Prospects for Nuclear Power in the UK', Nuclear Review Report, Cmnd 2860, HMSO, London, May.

Nuttall, W. (2005) *Nuclear Renaissance*, IOP Publishing, Bristol.

Ohya, Y., and Karasudani, T. (2010) 'A Shrouded Wind Turbine Generating High Output Power with Wind-lens Technology', *Energies* 3(4), pp. 634–649. http://www.mdpi.com/1996-1073/3/4/634

Onishi, N., and Fackler, M. (2011) 'Utility Reform Eluding Japan after Nuclear Plant Disaster', *New York Times*, 17 November. http://

DOI: 10.1057/9781137274335

www.nytimes.com/2011/11/18/world/asia/after-fukushima-fighting-the-power-of-tepco.html?_r=1&smid=tw-nytimes&seid=auto

Patel, T. (2012) 'EDF Wins Reprieve As Hollande Cools on Greens Nuclear Pact', Bloomberg news service, 25 April. http://www.bloomberg.com/news/2012-04-25/edf-wins-reprieve-as-hollande-cools-on-greens-nuclear-pact-1-.html

PEW (2011) 'Opposition to Nuclear Power Rises amid Japanese Crisis', PEW Research Centre, Washington DC, 21 March. http://pewresearch.org/pubs/1934/support-nuclear-power-japan-gas-prices-offshore-oil-gas-drilling

Pidgeon, N. (2008) 'Understanding Risk', Cardiff University report.

Pidgeon, N. (2011) 'Mind over Matter: Public Opinion and the Climate and Energy Debate', Environmental Research Web, 16 September. http://environmentalresearchweb.org/cws/article/opinion/47237

Pidgeon, N., Henwood, K., Parkhill, K., Venables, D., and Simmons, P. (2008a) 'Living with Nuclear Power in Britain: A Mixed Methods Study', School of Psychology, Cardiff University.

Pidgeon, N., Lorenzoni, I., and Poortinga, W. (2008b) 'Climate Change or Nuclear Power – No Thanks! A Quantitative Study of Public Perceptions and Risk Framing in Britain', *Global Environmental Change* 18, pp. 69–85.

PIRC (2010) 'Offshore Evaluation', Offshore Valuation Group, led by the Public Interest Research Centre, Machynlleth. http://www.offshorevaluation.org

PIU (2002) 'The Energy Review', Cabinet Office, Performance and Innovation Unit, London, February. http://webarchive.nationalarchives.gov.uk/+/http://www.cabinetoffice.gov.uk/strategy/work_areas/energy.aspx

POST (2007) 'Public Opinion on Electricity Options', Parliamentary Office of Science and Technology, POST Note 294, October. http://www.parliament.uk/business/publications/research/briefing-papers/POST-PN-294

Priestly, R. (2012) 'How Dangerous Is Fukushima?' *New Zealand Listener*, 27 February. http://www.listener.co.nz/current-affairs/science/japans-fukushima-plant-how-dangerous-is-it/

PWC (2010) 'A Roadmap to 2050 for Europe and North Africa', PriceWaterhouse Coopers, London. http://www.pwc.co.uk/eng/publications/100_percent_renewable_electricity.html

DOI: 10.1057/9781137274335

Radowitz, B. (2011) 'Germany Plans Faster Nuclear Exit', *Wall Street Journal*, 12 April. http://online.wsj.com/article/SB10001424052748703 518704576258280279913962.html

Reuters (2011a) 'German Utilities Group Favours 2020 Nuclear Exit', Reuters news service, 8 April. http://af.reuters.com/article/ energyOilNews/idAFLDE7371TU20110408

Reuters (2011b) 'Germany to Phase Out Nuclear Power: Deputy Minister', Reuters news service, 4 April. http://www.reuters. com/article/2011/04/04/us-germany-energy-nuclear-idUSTRE73330H20110404

Reuters (2011c) 'Olkiluoto 3 Nuke Plant May Be Delayed Further', Reuters news service, 12 October. http://af.reuters.com/article/ energyOilNews/idAFL5E7LC0M620111012

Reuters (2011d) 'UK Nuclear Builders Say Unfazed by Fukushima Delays', Reuters news service, 17 November. http://mobile.reuters. com/article/rbssEnergyNews/idUSL5E7MH2V920111117?irpc=932

Reuters (2012) 'RWE, E.ON Scrap 15 Billion Pound Government Nuclear Plan', Reuters news service, 29 March. http://uk.reuters.com/ article/2012/03/29/uk-eon-rwe-idUKBRE82S09720120329

REW (2012) 'Asia Report: After Quake, Japan Pushes for Asia Supergrid', Renewable Energy World, 12 March. http://www. renewableenergyworld.com/rea/news/article/2012/03/asia-report-after-quake-japan-pushes-for-asia-supergrid?cmpid=WNL-Wednesday-March14-2012

Rogers, S. (2011) 'Radiation Exposure: A Quick Guide to What Each Level Means', *Guardian* Datablog, 15 March. http://www.guardian. co.uk/news/datablog/2011/mar/15/radiation-exposure-levels-guide#_

Rose, J. (1967) *Automation: Its Uses and Consequences*, Oliver and Boyd, Edinburgh.

Roussely, F. (2010) Institute for Energy and Environmental Research translation of Roussely Report 'Future of the French Civilian Nuclear Industry'. http://www.psr.org/nuclear-bailout/resources/roussely-report-france-nuclear-epr.html

Safecast (2012) Safecast Maps website. http://maps.safecast.org/

Safegrounds (2011) 'Perspectives on the Health Risks from Low Levels of Ionising Radiation', Safegrounds Learning Network. http://www. safegrounds.com/radiation_risk.htm

Sayonara Campaign (2011) 'Goodbye to Nuclear Power Plants' campaign, Tokyo. http://sayonara-nukes.org/english/

DOI: 10.1057/9781137274335

Schaps, C. (2012) 'Germany Powers France in Cold Despite Nuclear U-Turn', Reuters news service, 14 February. http://af.reuters.com/article/energyOilNews/idAFL5E8DD87020120214?sp=true

Schiermeier, Q. (2011) 'Radiation Release Will Hit Marine Life', *Nature* 472, 12 April, pp. 145–146.

Scottish Government (2011) 'Routemap for Renewable Energy in Scotland 2011', Scottish Government, Edinburgh, June. http://www.scotland.gov.uk/Publications/2011/08/04110353/

Scottish Government (2012) 'Electricity Generation Policy Statement', Scottish Government, Edinburgh, March. http://www.scotland.gov.uk/Topics/Business-Industry/Energy/EGPS2012/DraftEPGS2012

Sekiguchi, T. (2012) 'Japan's Ex-Premier Turns Anti-Nuclear Activist', *Wall Street Journal*, 26 January. http://online.wsj.com/article/SB10001424052970204624204577180231906156286.html

Sermage-Faure, S., Laurier, D., Goujon-Bellec, D., Chartier, M., Goubin, A., Rudant, J., Hemon, D., and Clavel, J. (2012) 'Childhood Leukemia around French Nuclear Power Plants – The GeoCAP Study, 2002–2007', *International Journal of Cancer* 131(5), pp. E769–E780.

SGR (2011) letter of a formal complaint by the Nuclear Free Local Authorities Secretariat, Scientists for Global Responsibility, UK, and others concerning the BBC 2 'Fukushima – Is Nuclear Power Safe?' *Horizon* documentary. http://www.sgr.org.uk/resources/sgr-supports-joint-complaint-bbc-over-fukushima-documentary

SHE (1994) Gallup poll data cited in a submission to the UK Government's Nuclear Review by Stop Hinkley Expansion, September.

Shears, R. (2011) 'Scores of Schools in South Korea Closed over Fears of Radioactive Rain from Japan's Crippled Nuclear Plant', Mail Online, 7 April. http://www.dailymail.co.uk/news/article-1374369/Japan-nuclear-crisis-South-Korea-schools-closed-Fukushima-radiation-fears.html

Shirouzu, N., Dvorak, P., Hayashi, Y., and Morse, A. (2011) 'Bid to "Protect Assets" Slowed Reactor Fight', *Wall Street Journal*, 19 March. http://online.wsj.com/article/SB10001424052748704608504576207912642629904.html?mod=WSJAsia__LEFTTopStories

Smart, R. (2012) 'Handout Comes with Power Play', *The Times*, London, 14 February.

Smith School (2012) 'Towards a Low Carbon Pathway for the UK', Smith School of Enterprise and the Environment, University of Oxford, March.

Sorensen, B. (2011) *A History of Energy*, Earthscan, London.

Sourcewatch (2009) US electricity price comparisons service. http://www.sourcewatch.org/index. php?title=Comparative_electrical_generation_costs

Sovacool, B. (2008) 'Valuing the Greenhouse Gas Emissions from Nuclear Power: A Critical Survey', *Energy Policy* 36, pp. 2940–2953.

Sovacool, B. (2011) *Contesting the Future of Nuclear Power*, World Scientific, Singapore.

Sovacool, B., and Valentine, S. (2012) *The National Politics of Nuclear Power*, Routledge, London.

Spiegel Online (2011) 'Nuclear Phase Out', compilation of commentary, interviews and news reports, 5 April. http://www.spiegel.de/ international/germany/0,1518,755200-2,00.html

SRU (2011) 'Pathways Towards a 100% Renewable Electricity System', German Advisory Council on the Environment (SRU). http:// www.umweltrat.de/SharedDocs/Downloads/EN/02_Special_ Reports/2011_10_Special_Report_Pathways_renewables.html

Stirling, A. (2011) critique of *New Scientist* coverage of Fukushima, a redacted version of which was published by *New Scientist* on 13 April. http://www.newscientist.com/article/mg21028080.100-nuclear-futures.html

Surrey, J., and Huggett, C. (1976) 'Opposition to Nuclear Power: A Review of International Experience', *Energy Policy* 4, pp. 286–307.

Tabuchi, H. (2012) 'Japan Admits Nuclear Plant Still Poses Dangers', *New York Times*, 29 March. http://www.nytimes.com/2012/03/30/ world/asia/inquiry-suggests-worse-damage-at-japan-nuclear-plant. html?_r=2h

Tabuchi, H., Onishi, N., and Belson, K. (2011) 'Japan Extended Reactor's Life, Despite Warning', *New York Times*, 21 March. http://www.nytimes.com/2011/03/22/world/asia/22nuclear. html?scp=28&sq=Fukushima&st=cse

Taira, T., and Hatoyama, Y. (2011) 'Nationalize the Fukushima Daiichi Atomic Plant', *Nature* 480, 15 December, pp. 313–314. http://www. nature.com/nature/journal/v480/n7377/full/480313a.html

Takada, A. (2011) 'Japan's Food-Chain Threat Multiplies As Fukushima Radiation Spreads', Bloomberg news service, 25 July. http://www. bloomberg.com/news/2011-07-24/threat-to-japanese-food-chain-multiplies-as-cesium-contamination-spreads.html

DOI: 10.1057/9781137274335

Takano, T., and Takano, H. (2012) 'Fukushima: The Social Impact of a Nuclear Disaster', *Ecologist*, 15 February. http://www.theecologist. org/blogs_and_comments/commentators/other_comments/1240157/ fukushima_the_social_impact_of_a_nuclear_disaster.html

Tamman, M., Casselman, B., and Mozur, P. (2011). 'Scores of Reactors in Quake Zones', *Wall Street Journal*, 19 March. http://online .wsj.com/article/SB10001424052748703512404576208872161503008 .html

Telegraph (2011a) 'Japan Bans Fukushima Rice', *Daily Telegraph*, London, 17 November. http://www.telegraph.co.uk/news/worldnews/asia/ japan/8895903/Japan-bans-Fukushima-rice.html

Telegraph (2011b) 'Fukushima Caesium Leaks "Equal 168 Hiroshimas"', *Daily Telegraph*, London, 25 August. http://www.telegraph.co.uk/ news/worldnews/asia/japan/8722400/Fukushima-caesium-leaks- equal-168-Hiroshimas.html

Ten Hoeve, J., and Jacobson, M. (2012) 'Worldwide Health Effects of the Fukushima Daiichi Nuclear Accident', *Energy and Environmental Science*, advance article. http://pubs.rsc.org/en/content/ articlelanding/2012/ee/c2ee22019a

TEPCO (2012) summary of an interim internal analysis of the Fukushima incident, in English, TEPCO, January.

Teyssen, J. (2012) 'E.ON Confirms Strategy Focused on Renewables', *Windpower Monthly*, quoting Teyssen's comments to German business daily *Handelsblatt*, 30 March. http://www.windpowermonthly. com/go/windalert/article/1124937/?DCMP=EMC- CONWindpowerWeekly

Thomas, S. (2010) 'The EPR in Crisis', Public Services International Research Unit, University of Greenwich, London. http://216.250.243.12/EPRreport.html

Thomas, S. (2011) 'New Thinking Now Needed for Nuclear: Public Interest in Britain's Plans for New Nuclear Power Plants', Parliamentary Brief, 4 April.

Thomas, S. (2012) Preface to M. Cohen and A. McKillop, *The Doomsday Machine: The High Price of Nuclear Energy, the World's Most Dangerous Fuel*, Palgrave Macmillan, Basingstoke.

Toke, D. (2010) letter to the *Guardian*, London, 17 June. http://www. guardian.co.uk/environment/2010/jun/17/true-costs-nuclear- power?INTCMP=SRCH

DOI: 10.1057/9781137274335

Toke, D. (2012) 'Admission That Nuclear Is Less Cost-Effective Than Renewable', letter to the *Financial Times*, London, 28 April. http://www.ft.com/cms/s/0/04d8b5a6-8e00-11e1-b9ae-00144feab49a.html

Torello, A. (2012) 'Belgian Minister Still Seeking Path to Nuclear Phase-Out', Fox Business news service, 14 February.

TPOES (2008) 'Oil Crunch' report, from the Industry Task Force on Peak Oil and Energy Security, whose members included Arup, FirstGroup, Foster and Partners, Scottish and Southern Energy, Solarcentury, Stagecoach Group, Virgin Group and Yahoo! http://peakoiltaskforce.net

Trade and Industry Select Committee (1998) 'Conclusions of the Review of Energy Sources for Power Generation and the Government Response to the Fourth and Fifth Reports of the Trade and Industry Committee', House of Commons, London, Cmnd 4071, October.

UNDP/UNICEF (2002) 'The Human Consequences of the Chernobyl Nuclear Accident', United Nations Development programme/United Nations Children's Fund.

UNEP (2008) 'Green Jobs: Towards Decent Work in a Sustainable, Low-Carbon World', United Nations Environment Programme/ILO/IOE/ITUC, September. http://www.unep.org/labour_environment/features/greenjobs-report.asp

UNSCEAR (2000) 'Report by the UN Scientific Committee on the Effects of Atomic Radiation', New York and Vienna. http://www.unscear.org/unscear/en/chernobyl.html

UNSCEAR (2008) 'Sources and Effects of Ionizing Radiation', Vol. II, report to the General Assembly, with scientific annexes, by the United Nations Scientific Committee on the Effects of Atomic Radiation, New York and Vienna.

UNSCEAR (2011) report on Chernobyl, United Nations Scientific Committee on the Effects of Atomic Radiation, New York and Vienna, February. http://www.unscear.org/unscear/en/chernobyl.html

Wade, A. (2009) *Radiation and Reason*, self-published. http://www.radiationandreason.com/

Warry, R. (2011) 'Q&A: Health Effects of Radiation Exposure', BBC News – Health, 21 July. http://www.bbc.co.uk/news/health-12722435

WEC (2012) 'Nuclear Energy One Year after Fukushima', report by the World Energy Council, London, March.

DOI: 10.1057/9781137274335

Wicks, M. (2009) energy security report for the prime minister, Department of Energy and Climate Change, London.

Wikileaks (2011) Kano's views as reported in a leaked US embassy cable, relayed in the *Guardian*, London, 14 March. http://www.guardian. co.uk/world/us-embassy-cables-documents/175295

WNN (2009) 'U-Turn for Nuclear Opponents', World Nuclear News, 23 February. http://www.world-nuclear-news.org/newsarticle. aspx?id=24707

WNN (2011a) 'Chubu Agrees to Hamaoka Shut Down', World Nuclear News, 9 May. http://www.world-nuclear-news.org/newsarticle. aspx?id=30002&terms=hamaoka

WNN (2011b) 'Exposures and Progress at Fukushima Daiichi', World Nuclear News, 24 March. http://www.world-nuclear-news.org/ RS_Exposures_and_progress_at_Fukushima_Daiichi_240311.html

WNN (2011c) 'Maintain Nuclear Perspective, China Told', World Nuclear News, 11 January. http://www.world-nuclear-news.org/ NP_Maintain_nuclear_perspective_China_told_1101112.html

WNN (2011d) 'Growth Remains Nuclear's Future', World Nuclear News, 15 June. http://www.world-nuclear-news.org/NP_Growth_remains_ nuclears_future_1506111.html

WNN (2011e) 'New Nuclear Energy Policy for Taiwan', World Nuclear News, 3 November. http://www.world-nuclear-news.org/ NP-New_nuclear_energy_policy_for_Taiwan-0311117.html

WNN (2011f) 'US Public Remains Favourable to Nuclear', World Nuclear News coverage of Bisconti polls data, 24 February. http:// www.world-nuclear-news.org/NP-US_public_remains_favourable_ to_nuclear-2402114.html

WNN (2011g) 'Nuclear Still Cost Competitive in Japan, Study Says', World Nuclear News coverage of a study by Japan's Institute of Energy Economics of post-Fukushima nuclear costs, 2 September. http://www.world-nuclear-news.org/EE-Nuclear_still_cost_ competitive_in_Japan_study_says-0209114.html

WNN (2012a) 'Singh: Foreign Groups behind Anti-Nuclear Protest', World Nuclear News, 24 February. http://www. world-nuclear-news.org/NP_Singh_Foreign_groups_ behind_anti_nuclear_protest_2402121.html?utm_ source=World+Nuclear+News&utm_campaign=e6fb282e92- WNN_Daily_24_February_20122_24_2012&utm_medium=email

DOI: 10.1057/9781137274335

WNN (2012b) 'Extending Operating Lives of French Reactors Best Option', World Nuclear News coverage of French Auditors Court review of nuclear costs, 31 January. http://www.world-nuclear-news. org/NP-Extending_operating_lives_of_French_reactors_best_ option-3101124.html

WNN (2012c) 'IAEA Reviews Japan's Nuclear Restart Process', World Nuclear News, 31 January. http://www.world-nuclear-news.org/ RS_IAEA_reviews_Japans_nuclear_restart_process_3101121.html

WNN (2012d) 'Fukushima a "Temporary Blip" for UK Support', World Nuclear News coverage of Ipsos MORI December 2011 poll, 18 January. http://www.world-nuclear-news.org/NP-Fukushima_a_ temporary_blip_for_UK_support-1801124.html

WNN (2012e) 'Fukushima Impacts Global Nuclear Generation in 2011', World Nuclear News, 13 April. http://www.world-nuclear-news. org/EE-Fukushima_impacts_global_nuclear_generation_in_2011- 1304124.html

WNN (2012f) 'Optimism from Industry on Fukushima Anniversary', World Nuclear News, 9 March. http://www.world-nuclear- news.org/NP_Optimism_from_industry_on_Fukushima_ anniversary_0903121.html

WNN (2012g) 'Levy Nuclear Project Moved Back by Three Years', World Nuclear News, 2 May. http://www.world-nuclear-news.org/ NN_Levy_nuclear_project_moved_back_by_three_years_0205122. html

WNN (2012h) 'Approval for First Nuclear New Build in America', World Nuclear News, 9 February. http://www.world-nuclear-news.org/ RS_Approval_for_first_nuclear_new_build_in_America_0902121. html

WWF (2011a) 'Positive Energy: How Renewable Electricity Can Transform the UK by 2030', World Wide Fund for Nature, London. http://assets.wwf.org.uk/downloads/positive_energy_final_designed. pdf

WWF (2011b) 'The Energy Report – 100% Renewable Energy by 2050', World Wide Fund for Nature. http://www.wwf.org.uk/ research_centre/research_centre_results.cfm?uNewsID=4565

Yablokov, A., Nesterenko, V., Alexey, V., and Nesterenko, A. (2009) 'Chernobyl: Consequences of the Catastrophe for People and the Environment', *Annals of the New York Academy of Sciences*

DOI: 10.1057/9781137274335

1181, December. http://www.nyas.org/publications/annals/Detail.
aspx?cid=f3f3bd16-51ba-4d7b-a086-753f44b3bfc1

Yomiuri (2011) ' "Colossal blunder" on Radioactive Cattle Feed/Govt
Officials Admit Responsibility for Foul-Up That Let Tainted Beef
Enter Nation's Food Supply', *Daily Yomuiri*, 18 July. http://www.
yomiuri.co.jp/dy/national/T110717002520.htm

DOI: 10.1057/9781137274335

Index

Note: Page numbers in *italics* refer to tables.

DOI: 10.1057/9781137274335

DOI: 10.1057/9781137274335

DOI: 10.1057/9781137274335

DOI: 10.1057/9781137274335

DOI: 10.1057/9781137274335

DOI: 10.1057/9781137274335

CPSIA information can be obtained at www.ICGtesting.com
Printed in the USA
LVOW13*1636051213

364049LV00016B/798/P